CONJUGATED MICROPOROUS POLYMERS

Atsushi Nagai

JENNY STANFORD
PUBLISHING

Published by

Jenny Stanford Publishing Pte. Ltd.
101 Thomson Road
#06-01, United Square
Singapore 307591

Email: editorial@jennystanford.com
Web: www.jennystanford.com

British Library Cataloguing-in-Publication Data
A catalogue record for this book is available from the British Library.

Conjugated Microporous Polymers

ISBN 978-981-5129-66-3 (Hardcover)
ISBN 978-1-003-63652-6 (eBook)

Contents

Preface

The concept of porosity has existed in nature for thousands of years in structures such as rocks, biological tissues, and charcoal. The term ”porous” is usually used to describe materials that may be interconnected and have permanent voids that are permeable to liquids and gases. In modern times, porous materials are important in many areas of science and technology, and there has been an explosion in the number of new advanced porous functional materials in the past 20 years. The International Union of Pure and Applied Chemistry (IUPAC) classifies porous materials into three types based on the pore diameter (*d*): microporous ($d < 2$ nm), mesoporous (2 nm $< d <$ 50 nm), and macroporous ($d >$ 50 nm). Zeolite, an inorganic porous framework, is a staple of society with a large global market for a wide range of applications, from detergents to the removal of radioactive particles from nuclear waste. MOFs are extended, typically crystalline frameworks built of metal ions or clusters that provide directional binding to organic ligands through the preferential shape and coordination number of the metal, creating voids in the structure and resulting in porosity.

Both zeolites and MOFs incorporate metals, but pure organic porous materials are also available. In porous MOFs, metal ions coordinate with organic linkers, which are usually aromatic, to provide both rigidity and directionality. Porous organic materials also tend to be rigid and directionally bonded to prevent the collapse of the porous structure. This can be achieved by careful selection of the molecular building blocks. Not all porous materials need to be extended networks. For example, a porous organic cage (POC) is a discrete porous molecule in which the molecular cage is first synthesized and then assembled in a solid state in a separate step. Intrinsic microporous polymers (PIM) are a rare example of porosity in one-dimensional polymers, realized by inefficient packing of twisted polymer backbones.

In the past 20 years, a wide range of porous organic polymer networks have been introduced, including crystalline covalent organic frameworks (COF) and various amorphous networks, including hypercrosslinked polymers (HCP), covalent triazine frameworks (CTF), porous aromatic frameworks (PAF), and conjugated microporous polymers (CMP). With the exception of COF and a few CTFs, which are synthesized under thermodynamic control, other porous polymer networks are usually amorphous and disordered materials. These various materials are structurally different but held together by high porosity, lightweight elements, and strong two- or three-dimensional structures linked by covalent bonds.

When first discovered, CMP was unique among porous materials because it had extended π-conjugation throughout the porous three-dimensional network. PAFs are closely related to CMP and are formed using similar coupling chemistry, but do not have extended π-conjugation because they are linked by tetrahedral tetraphenylmethane nodes. Porous polymer network (PPN) is another term used to describe PAF. CTFs are also considered a subclass of CMPs because of their microporosity and extended π-conjugation, but they were developed separately and their formation chemistry was initially quite different.

In 2007, Prof. Andrew I. Cooper et al. reported the synthesis of microporous poly (aryleneethynylene) networks. This was the first example of what became to be known as CMP. Now, more than 10 years after its discovery, it is timely to reassess the status of the CMP. This book presents an updated overview of the CMP field, starting with an initial historical look at the origins of CMP and a list of major advances. Since CMP is an important platform for the development of new organic porous materials, various synthetic routes exist. Focusing on the utilization of optoelectronic properties such as photooxidation catalysis, gas storage, energy storage such as ion battery and supercapacitor, semiconductor, and photocatalytic H_2 production, the functional design of CMP has been discussed, which offers various potential applications.

Prof. Dr Atsushi Nagai
Ensemble3, Centre of Excellence

Chapter 1

Porous Organic Materials

Porous organic materials (POMs) are organic macromolecules that are considered emerging materials because of their high specific surface areas, tunable porosities, low densities, high chemical and thermal stabilities, variable composites, convenient post-functionalization, extended π-conjugations, and their high contents of carbons, nitrogen, oxygen, and other non-metallic atoms. POMs have been classified into four types: covalent triazine frameworks (CTFs), hyper-crosslinked polymers (HCPs), covalent organic frameworks (COFs), and conjugated microporous polymers (CMPs).

1.1 Porous Materials

Porous structures have existed in nature for a very long time, such as charcoal, biological tissues, and rocks [1]. The word "porous" is usually used to describe a material featuring interconnected permanent voids that allow the passage of gases or liquids [1]. Porous materials are applied in many technological and scientific fields, with a huge number of new advanced porous materials having been developed during the last two decades [2–14]. They can be broadly categorized into three different classes, as defined by IUPAC, in accordance with their pore. Macroporous materials have pore

Conjugated Microporous Polymers
Atsushi Nagai

ISBN 978-981-5129-66-3 (Hardcover), 978-1-003-63652-6 (eBook)
www.jennystanford.com

widths greater than 50 nm, mesoporous materials have pore widths between 2 and 50 nm, and microporous materials have pore widths less than 2 nm [15]. Many different synthetic porous materials have been produced. Zeolites are inorganic porous frameworks that are a staple product in society, having a global market with uses that range from detergents to the removal of radioactive particles from nuclear waste. Another class of inorganic-containing porous solids are hybrid inorganic-organic frameworks (MOFs), also known as porous coordination polymers (PCPs) [16–19]. MOFs are extended, typically crystalline frameworks constructed with metal ions or clusters that provide directional bonding to the organic ligand through the preferential geometry and coordination number of the metal, forming voids within the structure and giving rise to porosity.

Both zeolites and MOFs incorporate metals, but there is also a range of purely organic porous materials. In porous MOFs, MOFs are typically crystalline porous organic materials or porous coordination polymers obtained through the formation of coordinates bonds between organic ligands and clusters or metal ions to form voids within the framework structure, thereby resulting in porosity [20–22]. Crystalline organic porous materials also tend to be rigid and the bonding is somewhat directional to prevent collapse of the porous structure. This can be achieved by careful choice of the molecular building blocks. Not all porous materials need to be extended networks. For example, porous organic cages (POCs) are discrete porous molecules where the molecular cage is synthesized first and then assembled in the solid state in a separate step [23–25]. POC is a unique class of solution-processable molecular materials where porosity arises via the interconnection of intrinsic microporosity (PIMs), which is a rare example of porosity in a one-dimensional (1D) polymer, achieved through inefficient packing of the twisted polymer backbone [26–28].

Table 1.1 Example of porous organic polymers (POPs) networks

Covalent organic frameworks (COFs) Coupling chemistry: **Reversible**	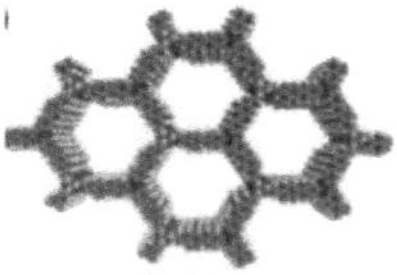 COFs

Covalent triazine frameworks (CTFs) Coupling chemistry: **Reversible cyclotrimerization**	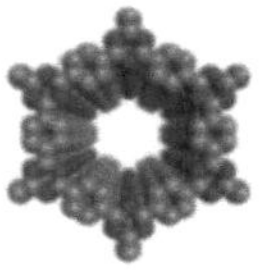 CTFs
Porous aromatic frameworks (PAFs) Coupling chemistry: **Irreversible C-C coupling**	 PAFs
Hyper-crosslinking polymers (HCPs) Coupling chemistry: **Irreversible Friedel Crafts Allylation**	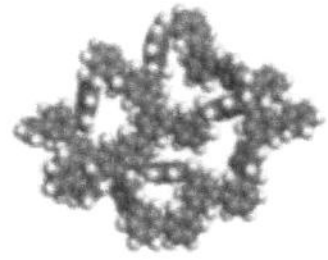 HCPs
Conjugated microporous polymers (CMP) Coupling Chemistry: **Irreversible C-C coupling or imine formation**	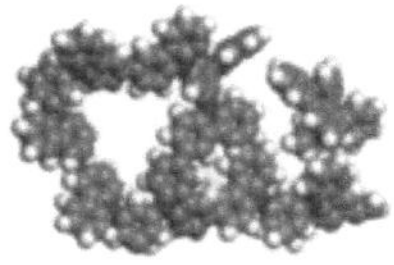 CMPs

In the past two decades, many groups have prepared various subclasses of POP frameworks, including crystalline covalent organic frameworks (COFs) [29–31], and various amorphous networks, such as porous triazine frameworks (CTFs) [32–34], porous aromatic frameworks (PAFs) [35–37], hyper-crosslinked polymers (HCPs) [38–40], and conjugated microporous polymers (CMPs) [41–43], as shown in Table 1.1

1.2 Porous Organic Polymers (POPs)

This section discusses the progress in the synthetic methodologies that are widely used for the preparation of four kinds of POPs: HCPs, CTFs, CMPs, and COFs.

1.2.1 Hyper-Crosslinking Polymers (HCPs)

HCPs are porous polymers that have attracted much attention for their extended conjugation; tunable porosities; high surface areas; high mechanical, chemical, and hydrothermal stabilities; variety of potential synthetic methods; and facile post-functionalization after HCP formation [44–49]. Many methods have been used to determine the chemical structures, thermal stabilities, morphologies, and textural characteristics, e.g., pore size diameters and surface areas, of HCPs, including Fourier transform infrared magnetic resonance (FTIR) spectroscopy, elemental analysis, solid-state nuclear magnetic resonance (NMR) spectroscopy, transmission electron microscopy (TEM), optical microscopy, field emission scanning electron microscopy (SEM), N_2 adsorption isotherms, and thermogravimetric analysis (TGA).

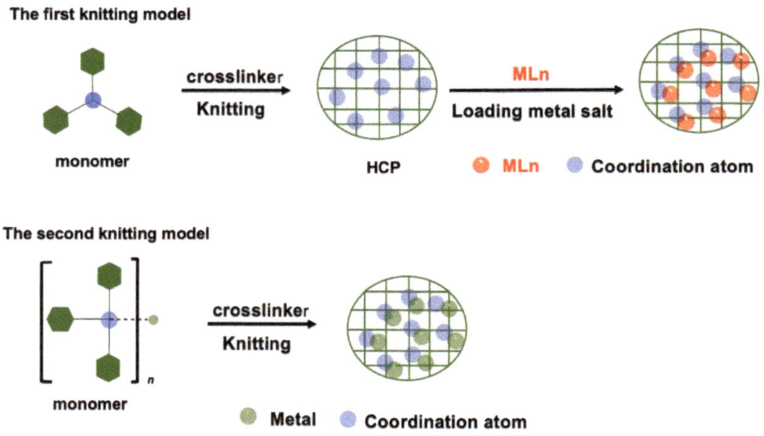

Figure 1.1 Schematic cartoon for incorporating metal species into HCP frameworks via two different knitting models.

Davankov and Tsyurupa were the first to generate a three-dimensional (3D) network structure of linear polystyrene (PS) with high-surface areas (up to 2000 $m^2\ g^{-1}$) as the first HCP materials, through post-crosslinking of a PS precursor in the swollen state [50]. The formation of HCPs can be achieved through acid-catalyzed Friedel-Crafts reactions involving a Lewis acid catalyst, a reaction solvent, and arene-based nucleophilic and electrophilic species [51].

To avoid products forming through side reactions during the knitting step, the reactions are performed under Ar and N_2 atmospheres to afford pure HCPs as solid materials [51]. The incorporation of metal species into HCPs can be accomplished via post-modification by introducing a metal salt or complex, or by using a metal complex as a comonomer or monomer during the knitting reaction (Figure 1.1) [51]. HCP materials have been employed in many applications, including energy storage; i.e., supercapacitor devices, and lithium, sodium, and potassium batteries; chemical sensing; heterogeneous catalysis; optoelectronic devices; drug delivery; gas capture and separation; and chromatographic separation [52–59].

1.2.2 Covalent Triazine Frameworks (CTFs)

CTFs are partially crystalline, amorphous porous materials that possess conjugated and layered structures [60]. Thomas and co-workers were the first to report examples of CTF materials containing aromatic rings and triazine moieties in the core [6]. CTFs have attracted much attention for their potential applications, such as CO_2 reduction, pollutant degradation, organic synthesis, water splitting, H_2O_2 and H_2 production, and organic semiconductor devices. [61–66]. Several approaches have been used for the preparation of CTFs, including cyclotrimerization of nitrile groups through direct or indirect methods in the presence of molten $ZnCl_2$ at 400 °C, superacid-catalyzed trimerization, Suzuki cross-coupling, Ni-catalyzed Yamamoto coupling, and P_2O_5 catalysis at 400 °C (Figure 1.2) [62]. For the preparation of monomers with planar characteristics, the use of non-planar monomers leads to the formation of amorphous or semicrystalline CTFs [67–70]. To date, only a few crystalline CTFs have been reported. For example, Tang, et al. found that the layered structures increased upon increasing the microwave power from 20 to 100 W [71]. Interestingly, Xu et al. prepared a two-dimensional (2D) CTF with a well-stacked structure within 10 min at 25 °C through a TfOH-catalyzed interfacial reaction, with simple purification procedures applied after the trimerization reactions (Figure 1.3) [72]. They found that the crystalline CTFs formed using this approach were AB-staggered; in contrast, mostly crystalline CTFs exhibiting the AA stacking mode were obtained when using the ionothermal method.

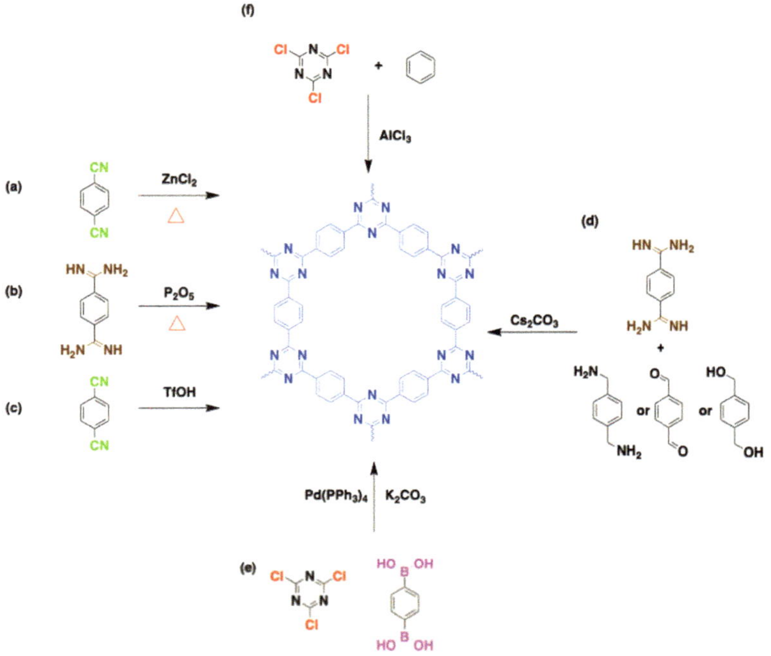

Figure 1.2 Formation of CTFs via direct or indirect approaches.

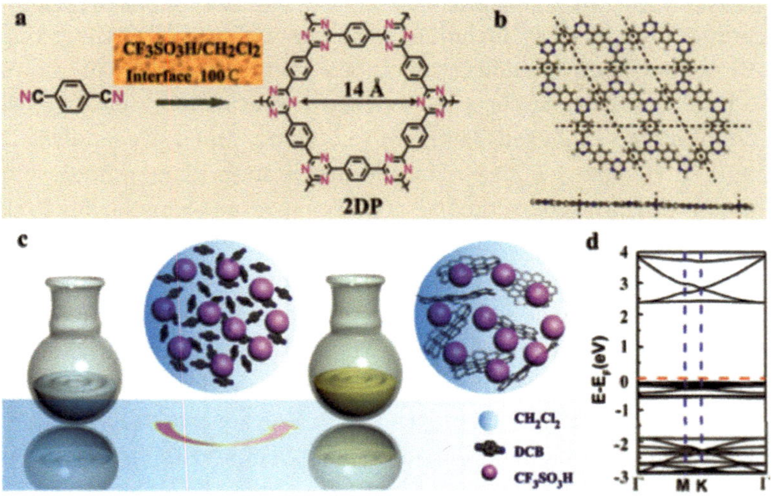

Figure 1.3 (a) and (c) Synthesis of 2D-CTFs via polymerization at CH_2Cl_2/CF_3SO_3H interfaces, (b) and (d) represent the atomic structure and DFT band structure of the 2D-CTF. Reproduced from ref. [72] with permission from American Chemical Society.

1.2.3 Covalent Organic Frameworks (COFs)

Covalent organic frameworks (COFs) are a class of microporous polymers that are distinguished by their crystallinity [73]. They are synthesized by linking together organic building blocks through dynamic covalent bond formation. Crystallinity in these materials because of the reversible nature of these reactions, which allows for some structural correction during synthesis resulting in a lower energy thermodynamic product, not too dissimilar to MOF synthesis. This contrasts with most other types of MOPs, which are the kinetic product of non-reversible covalent bond formation and are predominantly amorphous.

The first COF synthesis was reported in 2005 by Yaghi and co-workers [74], the result of the reversible formation of strong B-O bonds to form boroxine rings, which can be seen as analogous to the metal centers found in MOFs. The first COFs synthesized were named COF-1 and COF-5 with both being synthesized under solvothermal conditions in a dioxane/mesitylene mixed solvent. COF-1 is the simplest material in the series and is synthesized through the self-condensation of diboronic acid, which results in the formation of B3O3 rings connected by benzene rings in a hexagonal array. COF-5 is formed in a similar way but, instead of a self-condensation reaction, sees diboronic acid react with hexahydroxytriphenylene (HHTP) to form a five-membered BC2O2 ring bridged by benzene rings. In 2007, the COF family of materials was further extended by varying the strut length of the diboronic acid reagent (Figure 1.4) to yield COF-6 (1,3,5-benzenetriboronic acid), COF-8 (1,3,5-benenzetris(4-phenylboronic acid)), and COF-10 (4,4-biphenyldiboronic acid) [75].

All COFs synthesized were expected to form planar 2D organic sheets in which two distinct packing arrangements would form. A staggered AB arrangement where three-connected vertices lie over the center of the six-membered rings of neighboring layers, similar to that observed in graphite [77], or an eclipsed arrangement in which the sheets lay over one another as observed in boron nitrile [78]. Powder X-ray diffraction (PXRD) studies revealed distinct peaks directly related to each packing configuration which confirmed that COF-1 adopts the graphite-like configuration and each of the other COFs adopts the boron nitrile packing arrangement.

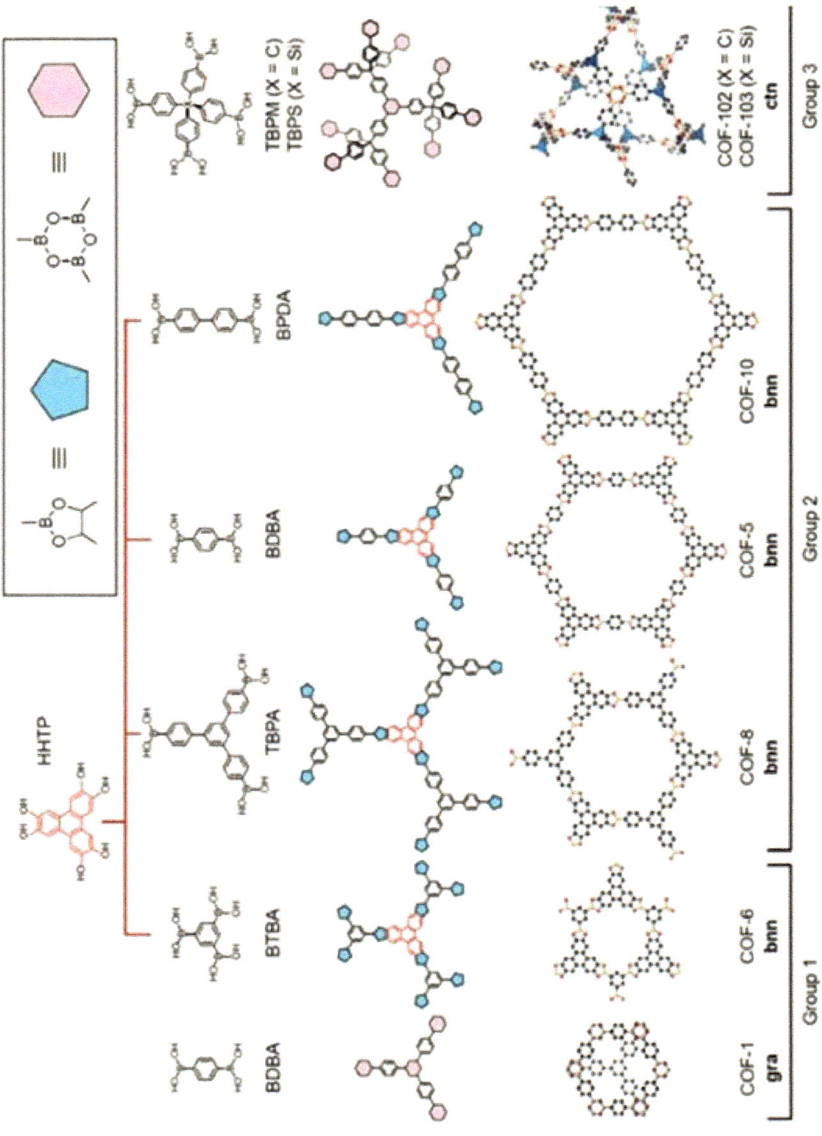

Figure 1.4 Selected COFs synthesized by the Yaghi group [76].

Up until 2007, all reported COFs were 2D due to their synthesis involving the use of planar monomers. This changed in 2007 when Yaghi and co-workers reported the first examples of 3D COFs [79]. These materials still involved reversible B-O bond formation to yield a crystalline product but employed 3D reagents in the form of tetra(4-hydroxyborylphenyl)methane and its silicon analog to yield COF-102 and COF-103, respectively. Likewise, COF-105 and COF-108 could be formed by reacting tetra(4-hydroxylborylphenyl)methane and its silicon analog via co-condensation with HHTP, respectively. All these materials were found to be crystalline.

By forming 3D COFs, it was found that the surface areas of the formed product far exceeded those of the 2D materials. The porosity and surface area of each sample were investigated through gas sorption studies. COF-1 and COF-5 both displayed type I isotherms and were found to have a BET surface area of 711 m^2/g and 1590 m^2/g, respectively. The larger surface area of COF-5 is a result of the larger mesopores in this material that are not present in COF-1. Conversely, COF-102 and COF-103 have much larger BET surface areas of 3472 m^2/g and 4210 m^2/g, respectively. More similar to those observed for other 3D materials such as ML-101 [80] and MIL-100 [81] – a result of the network propagating in three dimensions.

The high accessible surface area combined with the lower density framework COFs possess, when compared to MOFs, means these materials lend themselves very well to gas capture and storage applications. As such, over the years, COFs have been studied as sorbents for a number of different gases such as H_2, CH_4, and CO_2. COFs also display good stability under high pressure making them desirable for high-pressure gas uptake. For example, COF-102 and COF-103 are able to uptake over 1190 mg/g and 1200 mg/g of CO_2, respectively, at 55 bar and 298 K whilst still retaining their crystalline structure [76].

Other research groups have also taken an interest in COFs, owing to their lightweight frameworks, crystallinity, and high surface area. Work published by Lavigne and co-workers in 2008 showed how the microporosity in COFs could be tailored by the addition of alkyl chains onto the monomer units (Figure 1.5) [82]. This work is a great example of how the pore size of the material can be tuned by varying the monomer used during synthesis. By increasing the alkyl chain length, the pore diameter of the material becomes smaller. The tailoring of the pore size in these materials can be a useful method to increase the selectivity of the material toward different gases, in this case, becoming more selective toward H_2 and N_2.

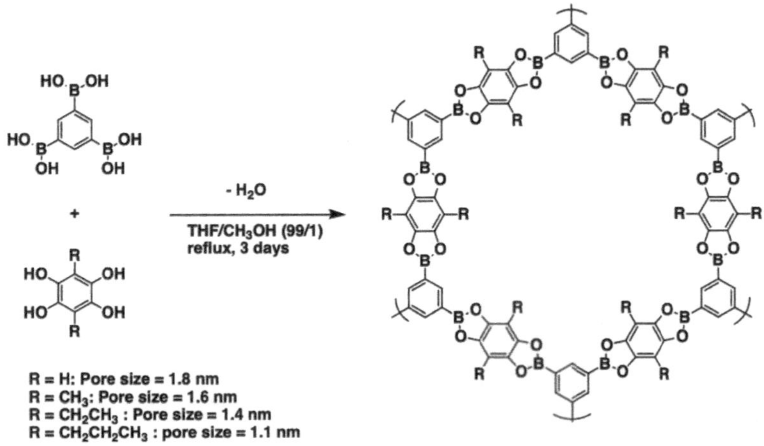

Figure 1.5 Synthesis of alkyl-modified COFs.

Up until 2009, all literature on COF synthesis involved the use of boronic acid-based monomers that formed a crystalline network because of reversible boroxine ring formation. However, this changed in 2009 when Yaghi and co-workers reported the first COF synthesized using imine linkage (Figure 1.6) named COF-300 [83]. This 3D network was synthesized as a result of the dehydration reaction between the tetrahedral building block tetra(4-anilyl) methane and linear linker terephthaldehyde. COF-300 was found to be crystalline, possessing a diamond-like structure. Gas sorption analysis revealed the sample had a BET surface area of 1360 m^2/g with pore size distribution analysis confirming the sample was micropores with a pose size centered around 7.2 Å, close to the 7.8 Å observed in the crystal model.

Over the last decade, there have been many more COF syntheses reported, which have used a variety of chemical linkages to yield crystalline and porous COF materials. These include imine linkages, hydrazone linkages, enamine linkages, and azine linkages as well as the traditional B-O and B-O-O bond-forming reactions. These different synthetic procedures have allowed the incorporation of different functionalities to be present within the final metal allowing these materials applied to many different applications. For example, in 2013, Jiang and co-workers synthesized Py-Azine-COF, which is an azine-linked COF capable of detecting trace amounts of the explosive compound picric acid [84]. In 2018, Ma and co-workers reported the synthesis of PyVg-COF, which included viologen units within the structure that ensure the 2D material does not stack upon itself. This means this material is very solution-stable and can be applied to optoelectronic device fabrication [85]. Further to this, COFs have been investigated for their applications in gas separation [76], wastewater treatment [86], chemosensing [87], and more [88].

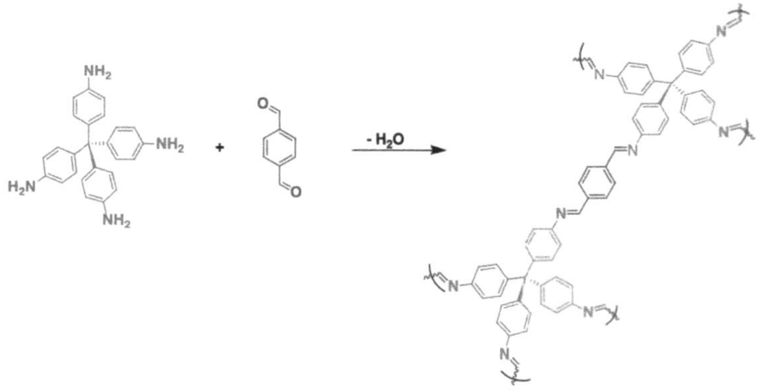

Figure 1.6 Synthesis of imine-linked COF-300.

Despite the fact that there has been much research in the field of COFs over almost decades, there are still numerous issues with these materials, which has limited their industrial applicability despite their excellent properties. Two main factors limiting COFs from realizing their potential are difficulties in synthesizing large amounts of material and poor processability. COFs require long reaction times (2–5 days) and are predominantly synthesized through solvothermal reactions carried out in seal tubes with low internal pressures. These two conditions are vital to ensure the final product is crystalline. There have recently been some reports of synthesizing COFs via other means which could allow for scalability such as microwave-assisted [89] and mechanochemical [90] synthesis methods. However, these too have their drawbacks and usually from materials that are less crystalline or have lower surface areas than their solvothermal analogs [91]. Likewise, even if it was possible to synthesize large quantities of a COF that met industrial requirements processability issues are still present and need to be overcome. This first COF thin film was not fabricated until 2017 [92], highlighting that much research needs to be carried out before COFs can realize their potential and be applied both industrially and commercially.

1.2.4 Conjugated Microporous Polymers (CMPs)

Conjugated microporous polymers (CMPs) are the only class of microporous organic polymers (MOP) to combine both a permanently porous structure alongside an extended fully π-conjugated network. Unlike COFs and to some extent CTFs, these materials are completely amorphous and show no crystallinity. This first CMP was reported in 2007 by the Cooper group and was the result of the palladium-catalyzed Sonogashira-Hagihara cross-coupling reaction between alkyne and halogen aromatic building blocks (Figure 1.7) [93]. CMP-1 was synthesized through the cross-linking reaction of 1,3,5-triethylnylbenzene and 1,4-diiodobenzene and was found to have a BET surface area of 834 m^2/g. Through amorphous, the pore size and surface area of CMPs can be tuned by varying the length of the organic linker, also known as the strut length. For example, CMP-3 possessed the longest strut length and consequently had the lowest BET surface area (522 m^2/g). This is due to the increased strut length imparting more flexibility into the materials, thereby lowering the free volume of the network. This was further corroborated in 2008 when the Cooper group reported CMP-0, which had the shortest strut length and as a result the highest BET surface area (1018 m^2/g) [94].

Homo-coupled CMP networks were reported in 2008 by Jiang et al. who successfully homo-coupling 1,3,5-triethynylbenzene and 1,4-diethynylbenzene to yield HCMP-1 and HCMP-2, respectively [95]. Both HCMP-1 and HCMP-2 were porous and had an apparent BET surface area of 842 m^2/g and 827 m^2/g, respectively. A later report by Li et al. showed how lithium ions could be incorporated into the framework and used to increase the capacity of the material compared to the unmodified HCMP-1. Such high uptake is owed in part to both the porous nature of the material as well as charge-induced dipole interaction between the Li^+ ions and H_2.

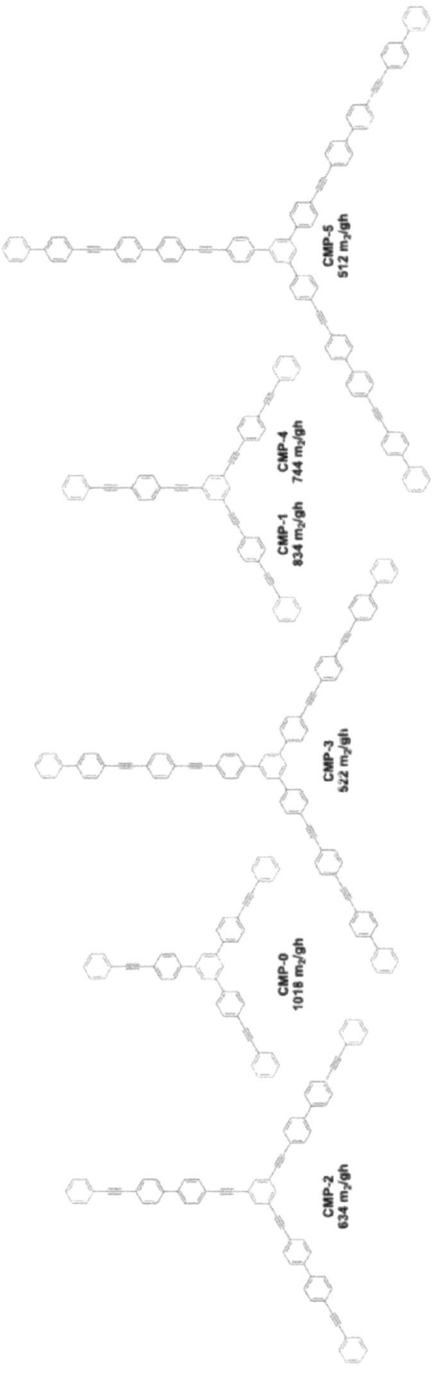

Figure 1.7 Schematic structures for a series of CMP networks.

A key advantage CMP synthesis has over other MOP synthesis, e.g., CTF synthesis, is that functional groups present on the monomers are present in the final materials. As such, this is a very convenient way to design materials, which can be tuned toward specific applications through judicious choice of starting materials. This was exploited by Dawson et al. to impart a degree of hydrophilicity into the final material [96]. The incorporation of hydroxyl groups into the CMP material leads to an increase in the network, as shown in 2011 by Jiang et al., who synthesized metal-containing CMP networks by cross-coupling 1,3,5-triethynylbenzene with 5,5-diboromo-2,2'-dipyridine [97]. These materials were mixed with a Re metal salt to incorporate a metal complex into the network bound by chelation to the bipyridine units. This material was able to act as a catalyst in the reductive amination of some disubstituted ketones.

Due to the extended conjugated network in which CMPs synthesize these materials can be applied to applications that exploit both their porosity but also properties that arise from the fully conjugated network. For example, CMPs have been applied toward photocatalytic hydrogen evolution, chemosensing, catalysis, light emission, and more with most applications exploiting the low band gap and optical properties of these materials. Though the scope of variation in CMP synthesis is wide and numerous functionalities can be introduced into the final materials leading to a number of applications, the final product is still completely insoluble, which limits the processability of the material.

In this chapter, porous organic materials (POMs) were introduced as organic macromolecules that are considered emerging materials because of their high specific surface areas, tunable porosities, low densities, high chemical and thermal stabilities, variable composites, convenient post-functionalization, and extended π-conjugations. The outline of the POMs showed specific properties and how to synthesize to classify into four types: covalent triazine frameworks (CTFs), hyper-crosslinked polymers (HCPs), covalent organic frameworks (COFs), and conjugated microporous polymers (CMPs). The next chapter describes the molecular design principles of CMPs and advancements in synthetic and structural studies.

References

1. Lee, J. S. and Cooper, A. I. (2020). Advances in Conjugated Microporous Polymers. *Chem. Rev.,* 120, pp. 2171–2214.
2. Cheng, Y. Z., Ding X., and Han, B. H. (2021). Porous Organic Polymers for Photocatalytic Carbon Dioxide Reduction. *ChemPhotoChem,* 5, pp. 406–417.
3. Wang, T. X., Liang, H. P., Anito, D. A., Ding X., and Han, B. H. (2020). Emerging Applications of Porous Organic Polymers in Visible-Light Photocatalysis. *J. Mater. Chem. A,* 8, pp. 7003. Materials Advances Review.
4. Amin, K., Ashraf, N., Mao, L., Faul C. F. J., and Wei, Z. (2021). Conjugated Microporous Polymers for Energy Storage: Recent Progress and Challenges. *Nano Energy,* 85, pp. 105958.
5. Mohamed, M. G., Tsai, M. Y., Su, W. C., EL-Mahdy, A. F. M. C., Wang, F., Huang, C. F., Dai, L., Chen T., and Kuo, S. W. (2020). Nitrogen-Doped Microporous Carbons Derived from Azo-Benzene and Nitrile-Functionalized Polybenzoxazines for CO_2 Uptake. *Mater. Today Commun.*, 24, pp. 101111.
6. Mohamed, M. G., Ebrahium, S. M., Hammam, A. S., Kuo S. W., and Aly, K. I. (2020). Enhanced CO_2 Capture in Nitrogen-Enriched Microporous Carbons Derived from Polybenzox-Azines Containing Azobenzene and Carboxylic Acid Units. *J. Polym. Res.*, 27, pp. 197.
7. Wang, S., Zhang, C., Shu, Y., Jiang, S., Xia, Q., Chen, L., Jin, S. Hussain, I., Cooper A. I., and Tan, B. (2017). Layered Microporous Polymers by Solvent Knitting Method. *Sci. Adv.*, 3, pp. e1602610.
8. M. M. Samy, G. Mohamed, T. H. Mansoure, T. S. Meng, Khan, M. A. R., Liaw, C. C., and Kuo, S. W. (2021). Solid State Chemical Transformations Through Ring-Opening Polymerization of Ferrocene-Based Conjugated Microporous Polymers in Host–Guest Complexes with Benzoxazine-Linked Cyclodextrin. *J. Taiwan Inst. Chem. Eng.*, DOI: 10.1016/j.jtice.2021.10.010.
9. Mohamed, M. G., Elsayed, M. H., Elewa, A. M., EL-Mahdy, A. F. M., Yang, C. H., Mohammed, A. A. K., Chou H. H., and Kuo, S. W. (2021). Pyrene-Containing Conjugated Organic Microporous Polymers for Photocatalytic Hydrogen Evolution from Water. *Catal. Sci. Technol.*, 11, pp. 2229–2241.
10. Li, X., Zhang, C., Cai, S., Lei, X., Altoe, V., Hong, F., Urban, J. J., Ciston, J. E., Chan M., and Liu, Y. (2018). Facile Transformation of Imine Covalent

Organic Frameworks into Ultrastable Crystalline Porous Aromatic Frameworks. *Nat. Commun.*, 9, pp. 2998.

11. Mohamed, M. G., Chen, W. C., El-Mahdy A. F. M., and Kuo, S. W. (2021). Porous Organic/Inorganic Polymers Based on Double-Decker Silsesquioxane for High-Performance Energy Storage. *J. Polym. Res.*, 28, pp. 219.
12. Samy, M. M., Mohamed, M. G., and Kuo, S. W. (2020). Directly Synthesized Nitrogen and Oxygen–Doped Microporous Carbons Derived from a Bioderived Polybenzoxazine Exhibiting High-Performance Supercapacitance and CO_2 Uptake. *Eur. Polym. J.*, 138, pp. 109954.
13. EL-Mahdy, A. F. M., Yu, T. C., Mohamed M. G., and Kuo, S. W. (2020). Secondary Structures of Polypeptide-Based Diblock Copolymers Influence the Microphase Separation of Templates for the Fabrication of Microporous Carbons. *Macromolecules*, 54, pp. 1030–1042.
14. Wu, J. Y., Mohamed M. G., and Kuo, S. W. (2017). Directly Synthesized Nitrogen-Doped Microporous Carbons from Polybenzoxazine Resins for Carbon Dioxide Capture. *Polym. Chem.*, 8, pp. 5481–5489.
15. Thommes, M., Kaneko, K., Neimark, A. V., Olivier, J. P., Rodriguez-Reinoso, F., Rouquerol, J., and Sing, K. S.W. (2015). Physisorption of Gases, with Special Reference to the Evaluation of Surface Area and Pore Size Distribution (IUPAC Technical Report). *Pure Appl. Chem.*, 87, pp. 1051.
16. Kondo, M., Yoshitomi, T., Matsuzaka, H., Kitagawa, S., and Seki, K. (1997). Three-Dimensional Framework with Channeling Cavities for Small Molecules: {[M2(4, 4′-bpy)3(NO3)4]·XH2O}$_n$ (M = Co, Ni, Zn). *Angew. Chem., Int. Ed. Engl.*, 36, pp. 1725–1727.
17. Long, J. R. and Yaghi, O. M. (2009). The Pervasive Chemistry of Metal-Organic Frameworks. *Chem. Soc. Rev.*, 38, pp. 1213–1214.
18. O'Keeffe, M. (2009). Design of MOFs and Intellectual Content in Reticular Chemistry: A Personal View. *Chem. Soc. Rev.*, 38, pp. 1215–1217.
19. Furukawa, H., Cordova, K. E., O'Keeffe, M., and Yaghi, O. M. (2013). The Chemistry and Applications of Metal-Organic Frameworks. *Science*, 341, pp. 1230444.
20. O'Keeffe, M. (2009). Design of MOFs and Intellectual Content in Reticular Chemistry: A Personal View. *Chem. Soc. Rev.*, 38, pp. 1215–1217.
21. Furukawa, H., Cordova, K. E., O'Keeffe, M., and Yaghi, O. M. (2013). The Chemistry and Applications of Metal-Organic Frameworks. *Science*, 341, pp. 1230444.

22. Li, L., Zhang, M., Li, R., Jiang, H., and Liu, Z. (2021). Facile Synthesis of Highly Luminescent Rod-Like Terbium-Based Metal–Organic Frameworks for Sensitive Detection of Olaquindox. *Anal. Methods,* 13, pp. 3785–3791.

23. Holst, J. R., Trewin, A. and Cooper, A. I. (2010). Porous Organic Molecules. *Nat. Chem.,* 2, pp. 915–920.

24. Jones, J. T. A., Hasell, T., Wu, X., Bacsa, J., Jelfs, K. E., Schmidtmann, M., Chong, S. Y., Adams, D. J., Trewin, A., and Schiffman, F., et al. (2011). Modular and Predictable Assembly of Porous Organic Molecular Crystals. *Nature,* 474, pp. 367–371.

25. Hasell, T. and Cooper, A. I. (2016). Porous Organic Cages: Soluble, Modular and Molecular Pores. *Nat. Rev. Mater.,* 1, pp. 16053.

26. Budd, P. M., Ghanem, B. S., Makhseed, S., McKeown, N. B., Msayib, K. J., and Tattershall, C. E. (2004). Polymers of Intrinsic Microporosity (PIMs): Robust, Solution-Processable, Organic Nanoporous Materials. *Chem. Commun.*, 0, pp. 230–231.

27. McKeown, N. B. and Budd, P. M. (2006). Polymers of Intrinsic Microporosity (PIMs): Organic Materials for Membrane Separations, Heterogeneous Catalysis and Hydrogen Storage. *Chem. Soc. Rev.,* 35, pp. 675–683.

28. McKeown, N. B. and Budd, P. M. (2010). Exploitation of Intrinsic Microporosity in Polymer-Based Materials. *Macromolecules,* 43, pp. 5163–5176.

29. Côte', A. P., Benin, A. I., Ockwig, N. W., Keeffe, M., Matzger, A. J., Yaghi, O. M. (2005). Porous, Crystalline, Covalent Organic Frameworks. *Science,* 310, pp. 1166–1170.

30. Ding, S.-Y. and Wang, W. (2013). Covalent Organic Frameworks (COFs): From Design to Applications. *Chem. Soc. Rev.*, 42, pp. 548–568.

31. Huang, N., Wang, P., and Jiang, D. (2016). Covalent Organic Frameworks: A Materials Platform for Structural and Functional Designs. *Nat. Rev. Mater.*, 1, pp. 16068.

32. Kuhn, P., Antonietti, M., and Thomas, A. (2008). Porous, Covalent Triazine-Based Frameworks Prepared by Ionothermal Synthesis. *Angew. Chem., Int. Ed.*, 47, pp. 3450–3453.

33. Ren, S., Bojdys, M. J., Dawson, R., Laybourn, A., Khimyak, Y. Z., Adams, D. J., and Cooper, A. I. (2012). Porous, Fluorescent, Covalent Triazine-Based Frameworks Via Room-Temperature and Microwave-Assisted Synthesis. *Adv. Mater.,* 24, pp. 2357–2361.

34. Katekomol, P., Roeser, J., Bojdys, M., Weber, J., and Thomas, A. (2013). Covalent Triazine Frameworks Prepared from 1,3,5-Tricyanobenzene. *Chem. Mater.,* 25, pp. 1542–1548.

35. Ben, T., Ren, H., Ma, S., Cao, D., Lan, J., Jing, X., Wang, W., Xu, J., Deng, F., Simmons, J. M., et al. (2009). Targeted Synthesis of a Porous Aromatic Framework with High Stability and Exceptionally High Surface Area. *Angew. Chem., Int. Ed.,* 48, pp. 9457–9460.

36. Ben, T. and Qiu, S. (2013) Porous Aromatic Frameworks: Synthesis, Structure and Functions. *CrystEngComm,* 15, pp. 17–26.

37. Konstas, K., Taylor, J. W., Thornton, A. W., Doherty, C. M., Lim, W. X., Bastow, T. J., Kennedy, D. F., Wood, C. D., Cox, B. J., Hill, J. M., et al. (2012). Lithiated Porous Aromatic Frameworks with Exceptional Gas Storage Capacity. *Angew. Chem., Int. Ed.*, 51, pp. 6639–6642.

38. Kuhn, P., Antonietti, M., and Thomas, A. (2008). Porous, Covalent Triazine-Based Frameworks Prepared by Ionothermal Synthesis. *Angew. Chem., Int. Ed.,* 47, pp. 3450–3453.

39. Ren, S., Bojdys, M. J., Dawson, R., Laybourn, A., Khimyak, Y. Z., Adams, D. J., and Cooper, A. I. (2012). Porous, Fluorescent, Covalent Triazine-Based Frameworks Via Room Temperature and Microwave-Assisted Synthesis. *Adv. Mater.,* 24, pp. 2357–2361.

40. Katekomol, P., Roeser, J., Bojdys, M., Weber, J., and Thomas, A. (2013). Covalent Triazine Frameworks Prepared from 1,3,5-Tricyanobenzene. *Chem. Mater.*, 25, pp. 1542–1548.

41. Jiang, J.-X., Su, F., Trewin, A., Wood, C. D., Campbell, N. L., Niu, H., Dickinson, C., Ganin, A. Y., Rosseinsky, M. J., Khimyak, Y. Z., et al. (2007). Conjugated Microporous Poly(Aryleneethynylene) Networks. *Angew. Chem., Int. Ed.,* 46, pp. 8574–8578.

42. Cooper, A. I. (2009). Conjugated Microporous Polymers. *Adv. Mater.,* 21, pp. 1291–1295.

43. Xu, Y., Jin, S., Xu, H., Nagai, A., and Jiang, D. (2013). Conjugated Microporous Polymers: Design, Synthesis and Application. *Chem. Soc. Rev.*, 42, pp. 8012–8031.

44. Huang, J. and Richard Turner, S., (2018). Hypercrosslinked Polymers: A Review. *Polym. Rev.,* 58, pp. 1–41.

45. Wei, Z., Chen, Q., and Liu, H., (2021). Hydroxyl Modified Hypercrosslinked Polymers: Targeting High Efficient Adsorption Separation Towards Aniline. *New J. Chem.*, 45, pp. 11607–11617.

46. Errahali, M., Gatti, G., Tei, L., Paul, G., Rolla, G., Canti, L., Fraccarollo, A., Cossi, M., Comotti, A., Sozzani, P., and Marchese, L. (2014). Microporous

Hyper-crosslinked Aromatic Polymers Designed for Methane and Carbon Dioxide Adsorption. *J. Phys. Chem. C*, 118, pp. 28699–28710.

47. Zhang, Y., Zhang, L., Zhang, X., Yang, D., Du, C., Wan, L., Au, C., Chen, J., and Xie, M. (2020). Pyridine-Based Hypercrosslinked Polymers as Support Materials for Palladium Photocatalysts and Their Application in Suzuki–Miyaura Coupling Reactions. *New J. Chem.*, 44, pp. 15202–15208.

48. Tan, L. and Tan, B. (2017). Hypercrosslinked Porous Polymer Materials: Design, Synthesis, and Applications. *Chem. Soc. Rev.*, 46, pp. 3322–3356.

49. Fontanals, N., Marce, R. M., Borrull F., and Cormack, P. A. G. (2015). Hypercrosslinked Materials: Preparation, Characterisation and Applications. *Polym. Chem.*, 6, pp. 7231–7244.

50. Davankov, V., Rogozhin, V., and Tsjurupa, M. (1973). US Pat. 3729457.

51. Gu, Y., Son, S. U., Li, T., and Tan, B. (2021). Low-Cost Hyper-crosslinked Polymers by Direct Knitting Strategy for Catalytic Applications. *Adv. Funct. Mater.*, 31, pp. 2008265.

52. J. Huang and S. Richard Turner. (2018). Hypercrosslinked Polymers: A Review. *Polym. Rev.*, 58, pp. 1–41.

53. Z. Wei, Q. Chen, and H. Liu. (2021). Hydroxyl Modified Hypercrosslinked Polymers: Targeting High Efficient Adsorption Separation Towards Aniline. *New J. Chem.*, 45, pp. 11607–11617.

54. M. Errahali, G. Gatti, L. Tei, G. Paul, G. Rolla, L. Canti, A. Fraccarollo, M. Cossi, A. Comotti, P. Sozzani, and L. Marchese. (2014). Microporous Hyper-crosslinked Aromatic Polymers Designed for Methane and Carbon Dioxide Adsorption. *J. Phys. Chem. C*, 118, pp. 28699–28710.

55. Y. Zhang, L. Zhang, X. Zhang, D. Yang, C. Du, L. Wan, C. Au, J. Chen, and M. Xie. (2020). Pyridine-Based Hypercrosslinked Polymers as Support Materials for Palladium Photocatalysts and Their Application in Suzuki–Miyaura Coupling Reactions. *New J. Chem.*, 44, pp. 15202–15208.

56. L. Tan and B. Tan. (2017). Hypercrosslinked Porous Polymer Materials: Design, Synthesis, and Applications. *Chem. Soc. Rev.*, 46, pp. 3322–3356.

57. N. Fontanals, R. M. Marce, F. Borrull, and P. A. G. Cormack. (2015). Hypercrosslinked Materials: Preparation, Characterisation and Applications. *Polym. Chem.*, 6, pp. 7231–7244.

58. V. Davankov, V. Rogozhin, and M. Tsjurupa. (1973). US Pat. 3729457.

59. Y. Gu, S. U. Son, T. Li, and B. Tan. (2021). Low-Cost Hypercrosslinked Polymers by Direct Knitting Strategy for Catalytic Applications. *Adv. Funct. Mater.*, 31, pp. 2008265.
60. Z. Qian, Z. J. Wang, and K. A. I. Zhang. (2021). Covalent Triazine Frameworks as Emerging Heterogeneous Photocatalysts. *Chem. Mater.*, 33, pp. 1909–1926.
61. P. Puthiaraj, Y. R. Lee, S. Zhang, and W. S. Ahn. (2016). Triazine-Based Covalent Organic Polymers: Design, Synthesis and Applications in Heterogeneous Catalysis. *J. Mater. Chem. A*, 4, pp. 16288–16311.
62. M. Liu, K. Jiang, X. Ding, S. Wang, C. Zhang, J. Liu, Z. Zhan, G. Cheng, B. Li, H. Chen, S. Jin, and B. Tan. (2019). Controlling Monomer Feeding Rate to Achieve Highly Crystalline Covalent Triazine Frameworks. *Adv. Mater.*, 31, pp. 1807865.
63. M. Liu, Q. Huang, S. Wang, Z. Li, B. Li, S. Jin, and B. Tan. (2018). Crystalline Covalent Triazine Frameworks by In Situ Oxidation of Alcohols to Aldehyde Monomers. *Angew. Chem., Int. Ed.*, 57, pp. 11968–11972.
64. X. Jiang, P. Wang, and J. Zhao. (2015). 2D Covalent Triazine Framework: A New Class of Organic Photocatalyst for Water Splitting. *J. Mater. Chem. A*, 3, pp. 7750–7758.
65. J. Li, P. Liu, H. Huang, Y. Li, Y. Tang, D. Mei, and C. Zhong. (2020). Metal-Free 2D/2D Black Phosphorus and Covalent Triazine Framework Heterostructure for CO_2 Photoreduction. *ACS Sustainable Chem. Eng.*, 8, pp. 5175–5183.
66. C. Xu, Q. Xie, W. Zhang, S. Xiong, C. Pan, J. Tang, and G. Yu. (2020). A Vinylene-Bridged Conjugated Covalent Triazine Polymer as a Visible-Light-Active Photocatalyst for Degradation of Methylene Blue. *Macromol. Rapid Commun.*, 41, pp. 2000006.
67. P. Katekomol, J. Roeser, M. Bojdys, J. Weber, and A. Thomas. (2013). Covalent Triazine Frameworks Prepared from 1,3,5-Tricyanobenzene. *Chem. Mater.*, 25, pp. 1542–1548.
68. 74 M. J. Bojdys, J. Jeromenok, A. Thomas, and M. Antonietti. (2010). Rational Extension of the Family of Layered, Covalent, Triazine-Based Frameworks with Regular Porosity. *Adv. Mater.*, 22, pp. 2202–2205.
69. P. Kuhn, A. Thomas, and M. Antonietti. (2009). Toward Tailorable Porous Organic Polymer Networks: A High-Temperature Dynamic Polymerization Scheme Based on Aromatic Nitriles. *Macromolecules*, 42, pp. 319–326.
70. W. Zhang, C. Li, Y. P. Yuan, L. G. Qiu, A. J. Xie, Y. H. Shen, and J. F. Zhu. (2010). Highly Energy- and Time-Efficient Synthesis of Porous Triazine-Based

Framework: Microwave- Enhanced Ionothermal Polymerization and Hydrogen Uptake. *J. Mater. Chem.*, 20, pp. 6413–6415.

71. J. Xie, S. A. Shevlin, Q. Ruan, S. J. A. Moniz, Y. Liu, X. Liu, Y. Li, C. C. Lau, Z. X. Guo, and J. Tang, J. (2018). Efficient Visible Light-Driven Water Oxidation and Proton Reduction by an Ordered Covalent Triazine-Based Framework. *Energy Environ. Sci.*, 11, pp. 1617–1624.

72. J. Liu, W. Zan, K. Li, Y. Yang, F. Bu, and Y. Xu. (2017). Solution Synthesis of Semiconducting Two-Dimensional Polymer via Trimerization of Carbonitrile. *J. Am. Chem. Soc.*, 139, pp. 11666–11669.

73. Nagai, A. (2020). *Covalent Organic Frameworks*, Jenny Stanford Publishing Pre. Ltd., ISBN 978-981-4800-87-7.

74. A. J. M. & O. M. Y. A. P. Cote, A. I. Benin, N. W. Ockwig, and M. O'Keeffee. (2005). *Science (80-.),* 310, pp. 1166–1170.

75. A. P. Côté, H. M. El-Kaderi, H. Furukawa, J. R. Hunt, and O. M. Yaghi. (2007). *J. Am. Chem. Soc.,* 129, pp. 12914–12915.

76. H. Furukawa and O. M. Yaghi. (2009). *J. Am. Chem. Soc.*, 131, pp. 8875–8883.

77. A. K. G. & K. S. Novoselov. (2007). *Nat. Mater.*, 6, pp. 183–191.

78. M. Xu, T. Liang, M. Shi, and H. Chen. (2013). *Chem. Rev.*, 113, pp. 3766–3798.

79. H. M. El-Kaderi, J. R. Hunt, J. L. Mendoza-Cortes, A. P. Cote, R. E. Taylor, M. O'Keeffe, and O. M. Yaghi. (2007). *Science (80-.),* 316, pp. 268–272.

80. C. Férey, C. Mellot-Draznieks, C. Serre, F. Millange, J. Dutour, S. Surblé, and I. Margiolaki. (2005). *Science (80-.),* 309, pp. 2040–2042.

81. P. Horcajada, C. Serre, M. Vallet-Regí, M. Sebban, F. Taulelle, and G. Férey. (2006). *Angew. Chemie - Int. Ed.,* 45, pp. 5974–597.

82. R. W. Tilford, S. J. Mugavero, P. J. Pellechia, and J. J. Lavigne. (2008). *Adv. Mater.*, 20, pp. 2741–2746.

83. F. J. Uribe-Romo, J. R. Hunt, H. Furukawa, C. Klöck, M. O'Keeffe, and O. M. Yaghi. (2009). *J. Am. Chem. Soc.,* 131, pp. 4570–4571.

84. S. Dalapati, S. Jin, J. Gao, Y. Xu, A. Nagai, and D. Jiang. (2013). *J. Am. Chem. Soc.,* 135, pp. 17310–17313.

85. L. Wang, C. Zeng, H. Xu, P. Yin, D. Chen, J. Deng, M. Li, N. Zheng, C. Gu, and Y. Ma. (2019). *Chem. Sci.,* 10, pp. 1023–1028.

86. J. Wang and S. Zhuang. (2019). *Coord. Chem. Rev.*, 400, pp. 213046.

87. F. Zamora, P. Albacete, A. E. Platero-Prats, E. M. Pérez, A. López-Moreno, and S. Mena-Hernando. (2018). *Chem. Commun.*, 55, pp. 1382–1385.

88. P. Wang, Q. Xu, Z. Li, W. Jiang, Q. Jiang, and D. Jiang. (2018). *Adv. Mater.*, 30, pp. 1801991.

89. N. L. Campbell, R. Clowes, L. K. Ritchie, and A. I. Cooper. (2009). *Chem. Mater.*, 21, pp. 204–206.

90. B. P. Biswal, S. Chandra, S. Kandambeth, B. Lukose, T. Heine, and R. Banerjee. (2013). *J. Am. Chem. Soc.*, 135, pp. 5328–5331.

91. S. Kandambeth, K. Dey, and R. Banerjee. (2019). *J. Am. Chem. Soc.*, 141, pp. 1807–1822.

92. K. Dey, M. Pal, K. C. Rout, S. S. Kunjattu, A. Das, R. Mukherjee, U. K. Kharul, and R. Banerjee. (2017). *J. Am. Chem. Soc.*, 139, pp. 13083–13091.

93. J. X. Jiang, F. Su, A. Trewin, C. D. Wood, N. L. Campbell, H. Niu, C. Dickinson, A. Y. Ganin, M. J. Rosseinsky, Y. Z. Khimyak, and A. I. Cooper. (2007). *Angew. Chemie - Int. Ed.*, 46, pp. 8574–8578.

94. J. X. Jiang, F. Su, H. Niu, C. D. Wood, N. L. Campbell, Y. Z. Khimyak, and A. I. Cooper. (2008). *Chem. Commun.*, 8, pp. 486–488.

95. A. Li, R. F. Lu, Y. Wang, X. Wang, K. L. Han, and W. Q. Deng. (2010). *Angew. Chemie - Int. Ed.*, 49, pp. 3330–3333.

96. R. Dawson, D. J. Adams, and A. I. Cooper. (2011). *Chem. Sci.*, 2, pp. 1173.

97. R. Dawson, A. Laybourn, Y. Z. Khimyak, D. J. Adams, and A. I. Cooper. (2010). *Macromolecules*, 43, pp. 8524–8530.

Chapter 2

Synthetic and Structural Studies of CMPs

Conjugated porous polymers (CMP) are a type of organic polymer with a π-conjugated backbone and permanent nanopores, which are porous materials in contrast to non-π-conjugated porous polymers and conventional conjugated polymers that are nanoporous. As a new material platform, CMP offers a high degree of freedom in the molecular design of conjugated frameworks and nanoholes. This chapter describes the molecular design principles, synthesis, and structural studies of CMP.

2.1 Design and Synthesis

2.1.1 Design Principles

The routes to design and synthesize CMP networks are full of challenges, which require researchers to have careful consideration for network structures and appropriate methods to include functionalities in CMPs. Generally, the network framework of CMP requires that the building blocks have at least two reactive groups. Although C-C cross-coupling of building blocks of different shapes is necessary to form a porous structure, self-condensation, which falls

Conjugated Microporous Polymers
Atsushi Nagai

ISBN 978-981-5129-66-3 (Hardcover), 978-1-003-63652-6 (eBook)
www.jennystanford.com

into the C_2, C_3, C_4, and C_6 categories, is effective for building block homocoupling, as shown in Figure 2.1. Self-condensation, such as $C_2 + C_2 + C_2$ [1, 2], $C_3 + C_3$ [3, 4], or $C_4 + C_4$ [5, 6], leads to the formation of three-dimensional networks. In the C-C reaction, on the other hand, combinations such as $C_2 + C_3$ [7, 8], $C_2 + C_4$ [9, 10], $C_2 + C_6$ [11], $C_3 + C_4$, $C_3 + C_6$, and $C_4 + C_6$ are common [12]. In both cases, the nanoholes are covalently anchored to a three-dimensional network, leaving permanent voids in the resulting CMP [12].

Thus, the establishment of CMP networks is driven by their skeleton architecture or potential applications with well-defined functional groups. Sometimes it is also necessary to address a task with the self-condensation of these factors together. It is well-known that the self-condensation of the homocoupling of single building blocks occurs earlier than that of CMPs, such as in the first success for the construction of COF-1, which was obtained by the self-condensation of 1,4-benzene diboronic acid and had crystalline structures and permanent porosity [13]. However, the discovery of such porous organic polymers is limited by their confined establishment of skeletons, which are merely the infinite duplication of homo-monomers. Thus, an important principle for designing functional CMPs, especially the bifunctional CMP, is that diverse building monomers with two or more active groups are necessary to endow the resultant CMP with accessible functionalities beyond their π-conjugated skeletons.

Therefore, a variety of building blocks for CMP synthesis have been developed, including arenes, condensed aromatic rings, phenylethynylene derivatives, heterocyclic units, and macrocyclic systems (Figure 2.1). Reactive groups such as bromoarenes, iodoarenes, aromatic boronic acids, cyano-substituted arenes, aromatic aldehydes, ethynyl-substituted arenes, and amino-substituted arenes are widely used in the synthesis of CMP. These groups give CMP different chemical structures and, as a result, perform other functions.

In order to obtain specific functional groups in a CMP network, it is important to have an appropriate method to induce functional groups in the CMP backbone. The most common methods are de novo synthesis to construct monomers and post-synthetic modifications of the resulting skeletons, which are also widely used to functionalize other porous materials such as MOF [14, 15] and

COF [16, 17]. Both methods are effective for CMP functionalization, but they employ different strategies to add functional groups and active sites to the CMP network. Post-synthesis modifications, on the other hand, coordinate the already synthesized CMP network. Post-synthesis modification is a well-known way to functionalize CMP because it is a versatile tool for creating variations of primitive CMP with the same spatial structure. This modification, also called post-synthesis functionalization, can achieve the diverse functions of CMP but can cause structural collapse of CMP. Therefore, to adjust the CMP network without compromising the structural stability of CMP, it is considered particularly important to intentionally consider the functionalization of CMP by post-synthesis modification. This method is typically employed when the desired CMP cannot be synthesized by constructing a functional monomer, which is difficult to obtain through conventional synthetic routes. Thus, the de novo synthesis and post-synthesis modification of CMP provide a decorative technique to tailor the properties of CMP networks and bring advanced applications to this porous organic polymer.

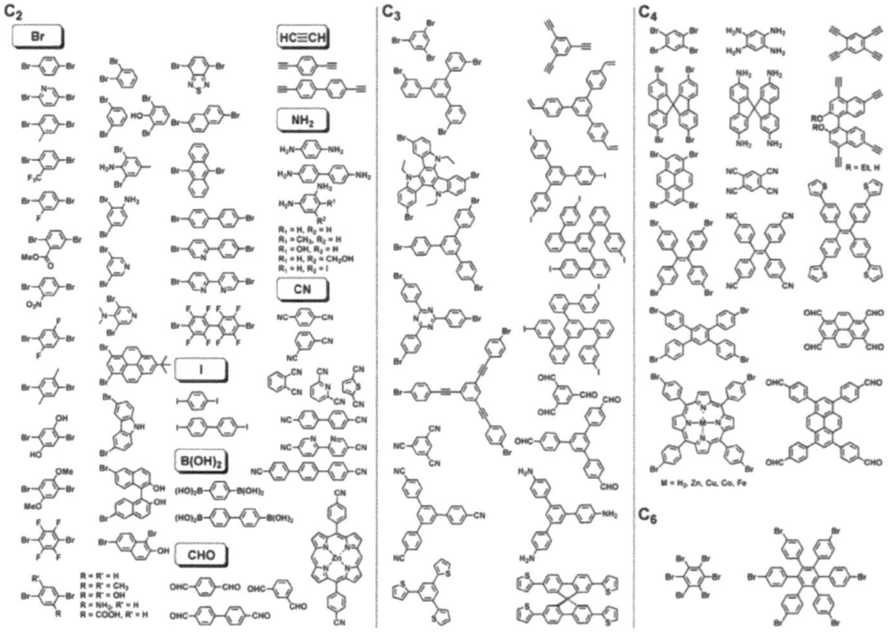

Figure 2.1 Schematic representation of the building blocks with different geometries for synthesis of CMPs. Adapted from [94].

2.2 Synthesis Routes

De novo synthesis: CMP networks can be formed by reacting two or more different monomers or, in some cases, by homocoupling a single monomer. The most common approach so far has been to combine the "core" of C_3 symmetry with the "linker" of C_2 symmetry, as proposed in the first CMP [18]. Homocoupling of trifunctional or multifunctional monomers can also result in three-dimensional porous networks. The most well-known example is PAF, which is formed from the homocoupling of a tetrahedral monomer with four functionalities [19)]. Two-dimensional monomers can obtain a three-dimensional polymer because the core-linker bond rotates freely during the reaction, and a complete two-dimensional pore may not form in subsequent reactions. The lack of reversibility of these kinetic bonds means that the structure cannot be modified into a thermodynamically more stable product (such as COF). Many three-dimensional building blocks, such as the tetrahedral monomers used in tetrahedral nodes, break the extended π-conjugation. However, some strategies work around this, such as the use of spirobifluorene linkers [20–22].

With MOF, the scope for CMP customization is nearly limitless. There is a wide range of options to adjust pore structure, morphology, and optoelectronics by changing the shape of the monomer or incorporating different heteroatoms or metals. It is also possible to modify CMP after synthesis [23, 24]. A variety of reactive coupling groups, including halogens, boronic acids, alkynes, alkenes, nitriles, amines, aldehydes, and activated phenol-substituted aromatic monomers, have been used for CMP synthesis and are expected to be discovered in the future. The following are typical reaction methods used in the de novo synthesis of CMP.

2.2.1 Sonogashira-Hagihara Coupling

The first was prepared using the Sonogashira-Hagihara cross-coupling (Scheme 2.1) [14], not because this reaction is uniquely effective, but because Cooper et al. wanted to introduce lightweight alkyne struts. This reaction couples alkyne-containing monomers with aryl halides in the presence of an amine base using palladium catalyst and copper co-catalyst, often tetrakis (triphenylphosphine) palladium and copper iodide. The presence of Cu co-catalyst is used for coupling due to its enhanced reactivity [25]. The copper-free Sonogashira-Hagihara reaction is also known [26], where four solvents were initially investigated: *N, N*-dimethylformamide (DMF), 1,4-dioxane, tetrahydrofuran (THF), and toluene. In general, CMPs formed in DMF were found to have the highest surface area and level of microporosity [27]. The Sonogashira-Hagiwara cross-coupling reaction can also be used for homocoupling reactions [27], since monomers containing alkynes may homocouple in the presence of oxygen [28]. Therefore, strict anaerobic and anhydrous conditions must be employed.

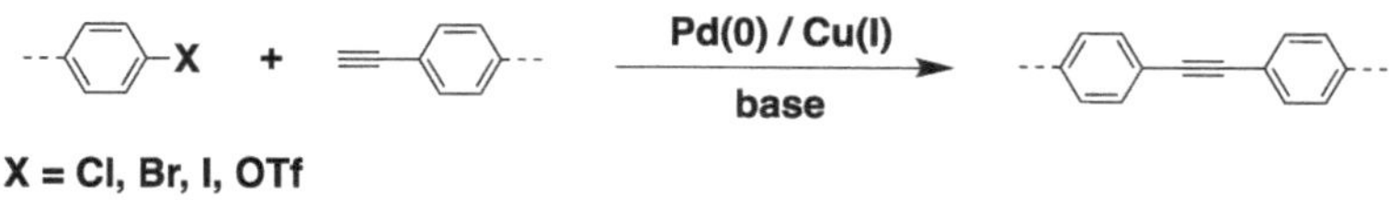

Scheme 2.1 Reaction scheme of Sonogashira-Hagihara cross-coupling.

2.2.2 Suzuki-Miyaura Coupling

The Suzuki-Miyaura cross-coupling reaction was discovered in 1979, as a way to attach aryl groups [29, 30]. This reaction uses a Pd (0) catalyst such as tetrakis (triphenylphosphine) -palladium (0) to combine arylboron monomers with aryl halides or sulfonates in DMF/water in the presence of a mild base such as K_2CO_3 [31]. This method has a variety of advantages, including commercial availability of boronic acid, wide compatibility of functional groups, and mild conditions, and therefore has the potential to scale up. However, the Suzuki-Miyaura cross-coupling (Scheme 2.2) is highly oxygen-sensitive and has the disadvantage of producing homocoupling and deboronized by-products. To solve this problem, the reaction needs to be deaerated sufficiently [32].

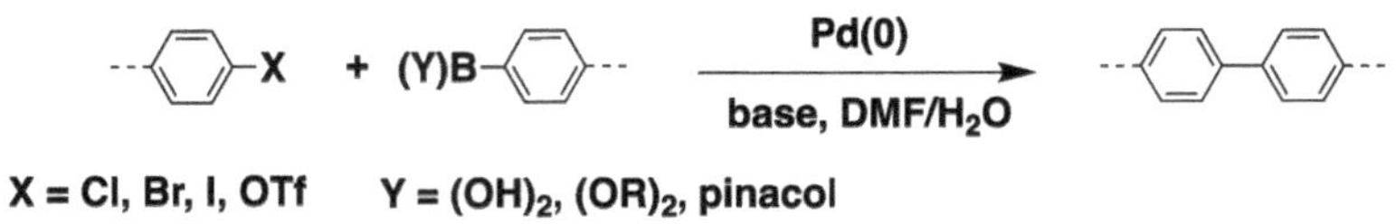

Scheme 2.2 Reaction scheme of Suzuki-Miyaura cross-coupling.

2.2.3 Yamamoto Coupling

Yamamoto coupling (Scheme 2.3) couples aryl halides with at least three reaction sites to bis (cyclooctadiene) nickel (0) (Ni $(cod)_2$) [33]. The advantage of this coupling method is that it requires a monomer with a single halogen functional group and requires only a single reaction procedure. Since aryl halide monomers are diverse, many porous networks can be designed by Yamamoto coupling involving CMP [34, 35]. When a three-dimensional monomer is used, the Yamamoto reaction can cause very inefficient packing and give a significantly higher surface area and porosity, as in the case of PAF-1 [36]. The drawback of this method is its sensitivity to water, which may hinder scale-up, and often requires reactions to be carried out in the glove box.

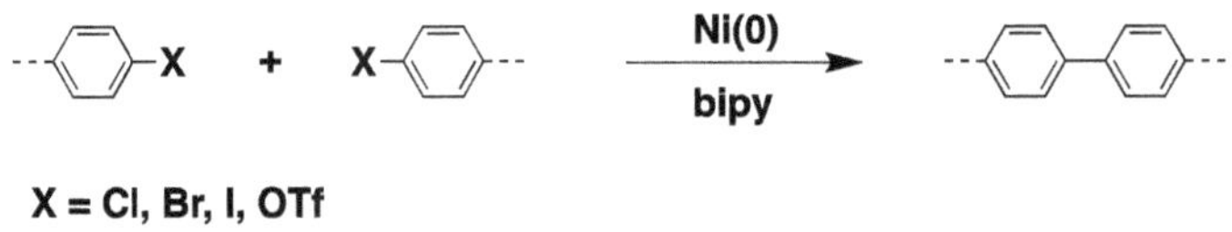

Scheme 2.3 Reaction scheme of Yamamoto coupling.

2.2.4 Heck Coupling

The Heck reaction, also called the Mirozoki-Heck reaction, couples an unsaturated halide with a primary alkene in the presence of a Pd catalyst and base to form a C = C bond between the two units (Scheme 2.4). CMP produced from this reaction is generally produced using Pd $(Ph_3)_4$, K_2CO_3 base, and DMF solvent under anaerobic conditions. The Suzuki-Heck one-pot reaction has also been used to form similar CMP products [37]. In this method, aryl halides are used as monomers, some of which are replaced by alkene functional groups at Suzuki-Miyaura, and then reacted with aryl halide carriers in the Heck reaction.

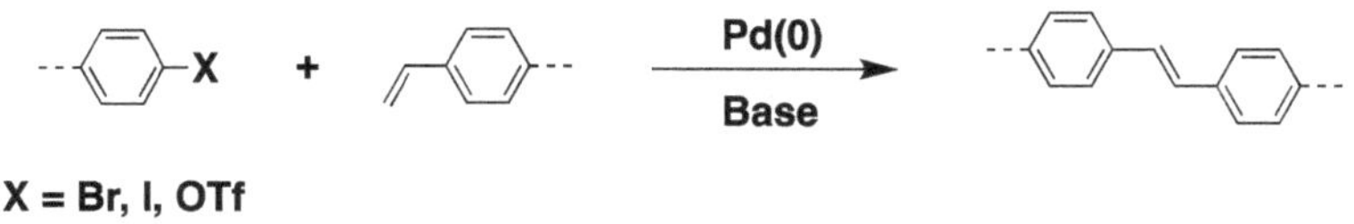

Scheme 2.4 Reaction scheme of Heck coupling.

2.2.5 Cyclotrimerization Reactions

Cyclotrimerization (Scheme 2.5) typically begins with an aromatic six-membered ring. For example, starting with three alkyne monomers, these are further connected to form an extended network [38]. Similarly, the use of a monomer with a cyano functional group results in the formation of a C_3N_3 *s*-triazine ring. Nitrile cyclization trimerization can be performed in molten $ZnCl_2$ under high-temperature conditions above 400 ºC [39–41]. However, under these harsh conditions, only the most stable monomers can be used. Therefore, a degree of carbonization was also developed to allow the reaction to proceed at moderate temperatures at room temperature or under microwave heating [42]. This method uses trifluoromethanesulfonic acid in $CHCl_3$ as a solvent, but can also occur in microwave synthesis without using a solvent. In the microwave-assisted reaction, a microwave power of 120–460 W can significantly reduce the reaction time by several tens of minutes. Recently, a low-temperature polycondensation approach was reported to produce triazines [43]. Trimerization in which different groups are introduced into monomers with cyano-functional groups has also been reported. The amide group has been shown to trimerize in the presence of P_2O_5 as a catalyst at temperatures between 350 and 550 ºC [44]. Unconventional covalent quinazoline networks that are fully aromatic and contain a tricycloquinazoline core were synthesized using o-aminonitrile monomers under ion temperature conditions using $ZnCl_2$ [45].

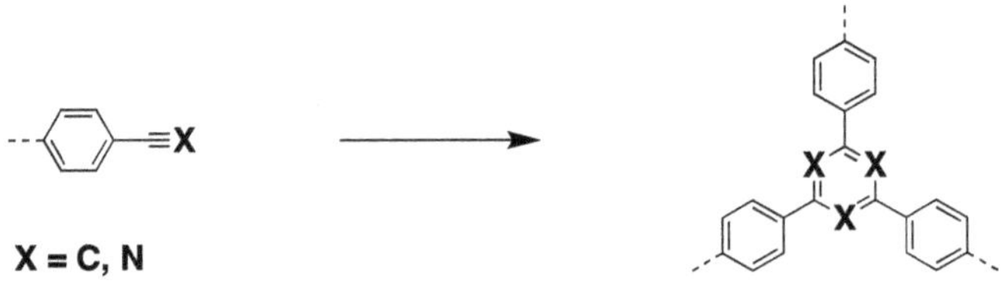

Scheme 2.5 Reaction scheme of cyclotrimerization.

2.2.6 Phenazine Ring Fusion

Ladder polymers have been manufactured via fusion of phenazine rings since 1966 [46, 47]. Aryl diamines and aryl diketones are connected by reacting certain high-boiling solvents, such as *N, N*-dimethylacetamide hexamethylphosphoramide, with 16% polyphosphate at a temperature of 250 ºC (Scheme 2.6). More recently, a fusion of phenazine rings has been used to produce Aza-CMP [48]. In this method, a 3D ladder CMP network is formed at high temperatures (300–500 ºC) in a vacuum ampoule in the presence of $AlCl_3$ using $C_2 + C_6$ monomers. Recently, a mild solvothermal method for preparing Aza-CMP was developed by refluxing the monomer in a 1:4 mixture of dioxane and acetic acid [49].

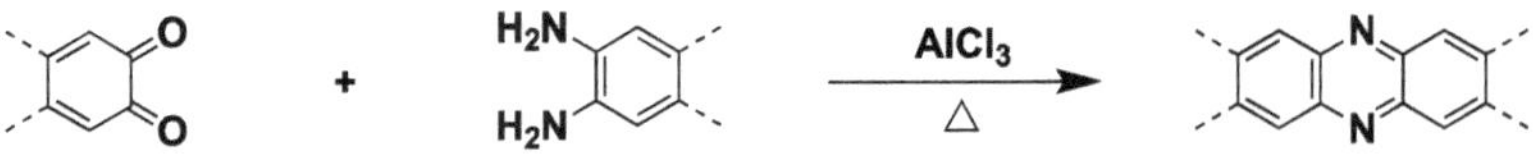

Scheme 2.6 Reaction scheme of phenazine ring fusion.

2.2.7 Schiff-Base Condensation

The Schiff base reaction has been used to synthesize both CMP and imine COF (Scheme 2.7) [50]. The reaction is metal-free, which saves money and ensures no metal remains in the CMP network. This reaction forms an imine bond from a monomer functionalized with an amine and an aldehyde [51]. The advantage of the Schiff base reaction is that it produces N-containing CMP, which is preferred for applications such as CO_2 selective adsorption [52, 53].

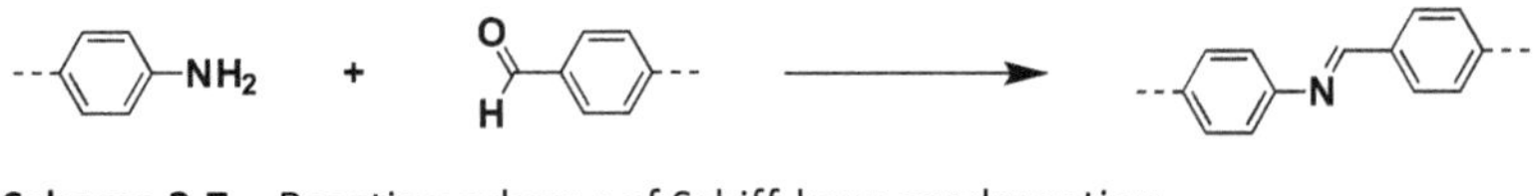

Scheme 2.7 Reaction scheme of Schiff-base condensation.

2.2.8 Heterocycle Linkages

Using monomers similar to those used for Schiff base condensation, benzimidazole bonds can be formed using ortho diamine in the presence of an aldehyde [54]. Rabbani and El-Kaderi carried out this reaction in DMF under inert conditions, followed by further reaction under air to produce a porous polymer. Thiazolothiazole-linked CMP can be synthesized using dithiooxamide and aldehyde monomers, reaction with DMF or nitrobenzene at 150 ºC [55], or solvothermal reaction in DMF at 160 ºC (Scheme 2.8) [56].

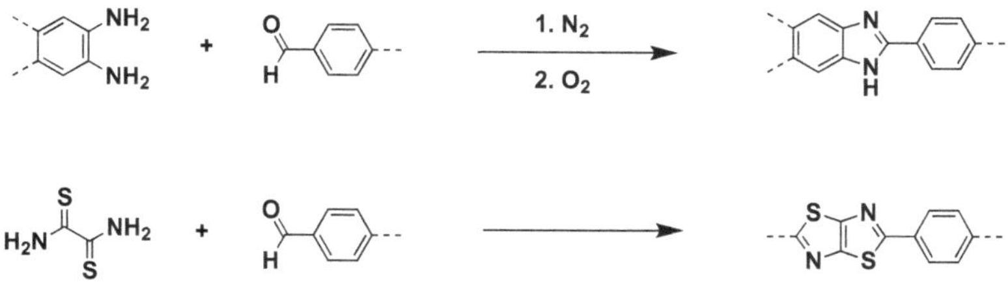

Scheme 2.8 Reaction scheme of heterocycle ring formation.

2.2.9 Alkyne Metathesis

In 1968, Bailey et al. reported the first alkyne metathesis reaction using a tungsten oxide catalyst on silica at 350 ºC (Scheme 2.9) [57]. Using this method, CMP was produced in $CHCl_3$ at a mild temperature using a Mo(VI)-based catalyst [58]. Although this reaction is reversible, it has been reported that the resulting CMP is semicrystalline. This reaction differs from the homocoupling of terminal alkynes in that only one alkyne group, not a diyne, exists in the product between the two linked monomers.

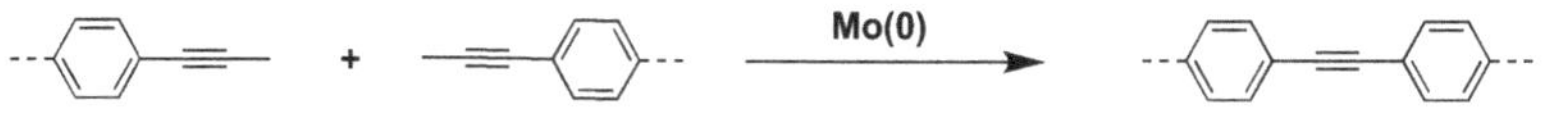

Scheme 2.9 Reaction scheme of alkyne metathesis.

2.2.10 Oxidative Coupling

Terminal alkynes can be combined in the presence of O_2 under Sonogashira-Hagiwara conditions (Pd (II)/Cu(I), Et_3N, solvent) [59]. CMP can also be produced by the oxidative Scholl reaction of electron-rich aromatics using Lewis acid catalysts such as $FeCl_3$ and $AlCl_3$ in a $CHCl_3$ solvent [60, 61]. This method, first observed by Friedel and Crafts more than 100 years ago [62], uses a relatively inexpensive catalyst and is therefore easier to manufacture and less costly (Scheme 2.10).

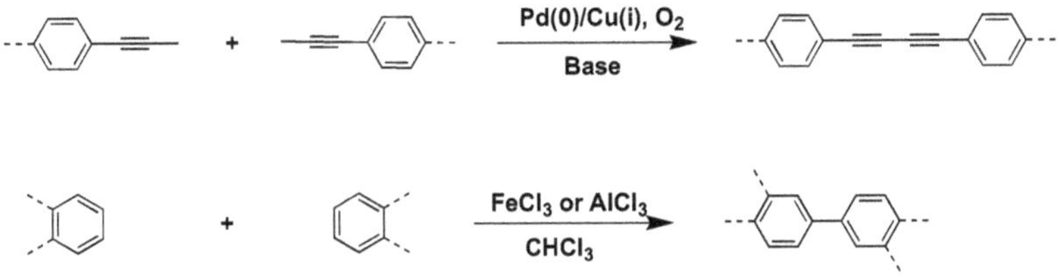

Scheme 2.10 Reaction scheme of oxidative coupling.

2.2.11 Buchwald-Hartwig Amination

Carbon-nitrogen bonds can be formed by a simple synthesis using a Pd catalyst and a base. It was recently used to form N-rich CMP with aryl halides and arylamines in toluene at 110 ºC in the presence of bis(dibenzylideneacetone)palladium(0) ($Pd(dlba)_2$), 2-dicyclohenylphosphino-2',4',6'-triisopropylbiphenyl (XPhos) and sodium tert-butoxide (NaOtBu) (Scheme 2.11) [63, 64].

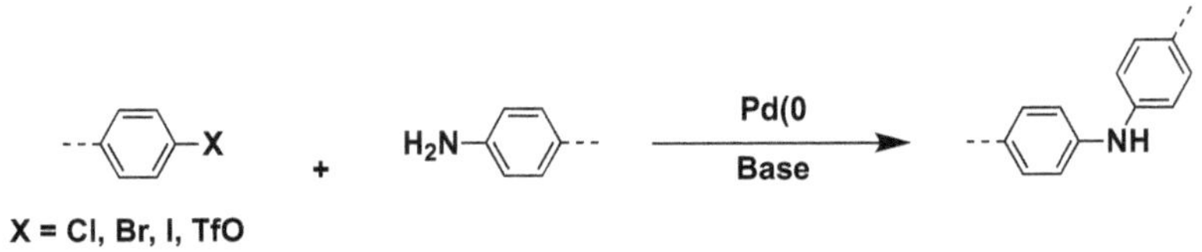

Scheme 2.11 Reaction scheme of Buchwald-Hartwig amination.

2.2.12 Electropolymerization

Oxidative coupling can also be carried out via an electrochemical route using cyclic voltammetry (CV) to obtain CMP films (Scheme 2.12) [65–68]. Ma et al. report the first synthesis of electropolymerized CMP in an acetonitrile/CH_2Cl_2 electrolyte using a carbazole functionalized monomer and Bu_4NPF_6 in a potential range of -0.8~0.97 V with a scan rate of 50 mV s^{-1}. The oxidation potential of the monomer was observed at 0.93 V, which was attributed to the oxidation of the carbazole group. Then, as the scan cycle increased, a new peak was observed at a lower potential due to the formation of carbazole dimers and the growth of the CMP film on the electrode [65].

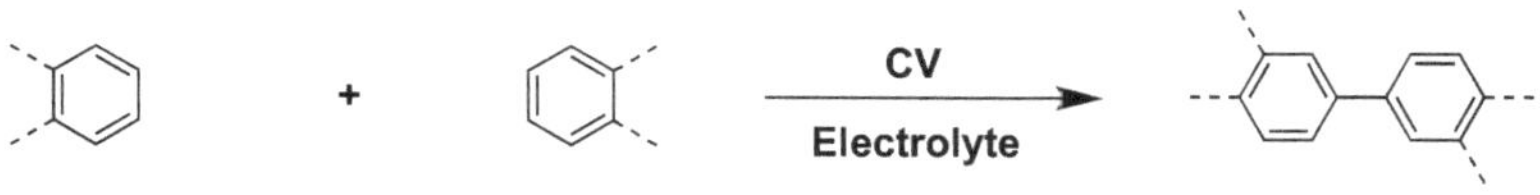

Scheme 2.12 Reaction scheme of electropolymerization.

2.2.13 Hypercrosslinking Linear Conjugation

An interesting approach to manufacturing CMP is to hyper-crosslink a preformed linear conjugated polymer. When linear polyanilines and linear conjugated polypyrroles were combined with cross-linker, the BET-specific surface area was increased to 632 m^2/g and 732 m^2/g, respectively [69, 70]. The method of reaction, the choice of solvent, and the choice of crosslinker greatly affect the porosity (Scheme 2.13).

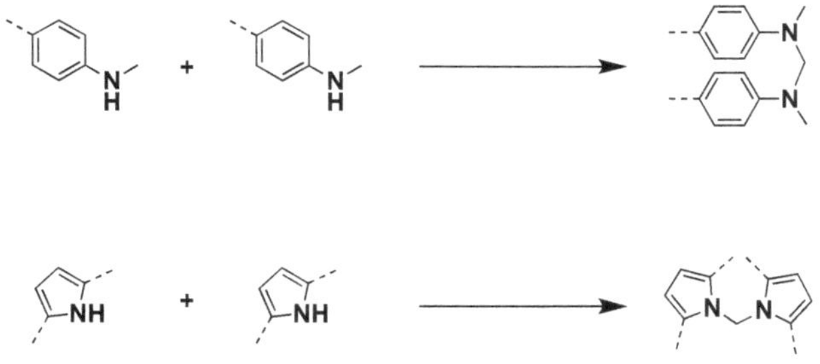

Scheme 2.13 Reaction scheme of hypercrosslinking linear conjugated polymers.

2.2.14 Post-Synthesis Modification

Post-synthesis modification (PSM) involves chemical modification after synthesis of a CMP network, compared to de novo synthesis, which builds CMP from the bottom up and thus guarantees the highest degree of grafting at the molecular level. However, the ease with which the benzene rings in many polymers can be functionalized makes it possible to achieve a very high degree of grafting by post-synthetic modification of CMP. In 1990, Hoskins and Robson originally proposed the concept of PSM [71], which was introduced to functionalize MOFs by Wang and Cohen until 2007 [72]. Subsequently, this strategy has been widely applied to functionalize all porous materials, including CMPs [73], POPs [74], and COFs [75], due to its convenience. PSM has proven to be a general, practical approach for the functionalization of porous materials after years of practical employment.

Metallization: The incorporation of active metal sites into the CMP network greatly introduces useful chemical functionality into the pores of the CMP. Several pioneering studies have been reported to obtain metal-organic conjugated porous polymers by loading metalloporphyrin cores onto CMP networks, greatly expanding the applications of CMP, which effectively combines chemical and physical properties in a single porous material [76]. Cooper et al. describe two versatile strategies for constructing metal-organic CMP (MO-CMP) without porphyrin-driven monomers (Figure 2.2) [77]. Compared to a typical MOF, a metal or metal cluster serves as a connecting node to establish an open metal-organic framework. As a result, MO-CMP metal sites are no longer needed to connect multiple organic monomers. This allows them to introduce active metal sites and various functionalities without breaking the skeleton. In addition, the extended p-conjugation of CMP can be maintained by modifying metal sites after CMP network synthesis. CMP-$(Bpy)_x$ was prepared by manipulating the content of active metal sites through Sonogashira-Hagiwara cross-coupling of 1,3,5-triethylbenzene with 5.5'-dibromo -2.2'-bipyrine and 1,4-dibromo-benzene at different ratios of bipyridine and aromatic. The resulting CMP-$(Bpy)_x$ was treated with a solution of $[Re(CO_3)Cl]$ toluene to incorporate Re metal sites, and the successful binding of Re to the network was verified by FT-IR spectra compared with bare CMP-$(Byp)_x$. Thus, Canivet and colleagues explored a family of bipyridine-based CMPs with different amounts of Ni metal sites for their potential catalytic performance in the C-H arylation of benzothiophene. The results showed that these Ni-doped CMPs with abundant active sites have high catalytic activity and a wide range of applied substrates [78]. Post-synthetic modifications can precisely tune the amount of metal loading in the CMP network and control the chemically active function on demand without compromising the extended π-conjugated structure, thereby maintaining the high porosity of CMPs and promising effective molecular transport in post-functional CMP.

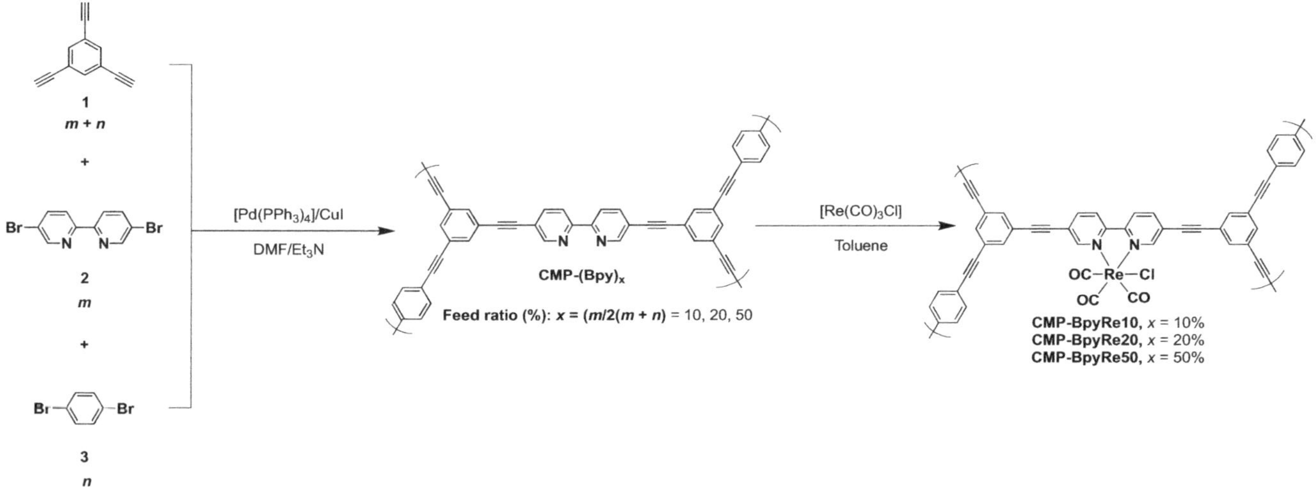

Figure 2.2 Preparation of metal-CMP via post-synthetic metalation.

Covalent Modification: Besides introducing various metals into the CMP network to construct dual-functional CMP materials, post-synthesis modification of CMP by covalent bonds is an important way to change the chemical and physical properties of CMP in its original form. Over the past few years, comprehensive research has been conducted on post-synthetic modifications of porous materials, and post-synthetic functional transformations are very useful for tailoring the pore environment of CMP for specific applications. Recently, Ma et al. successfully demonstrated a dual-functional CMP adsorbent for efficient uranium capture through post-synthesis functional conversion (Figure 2.3) [79].

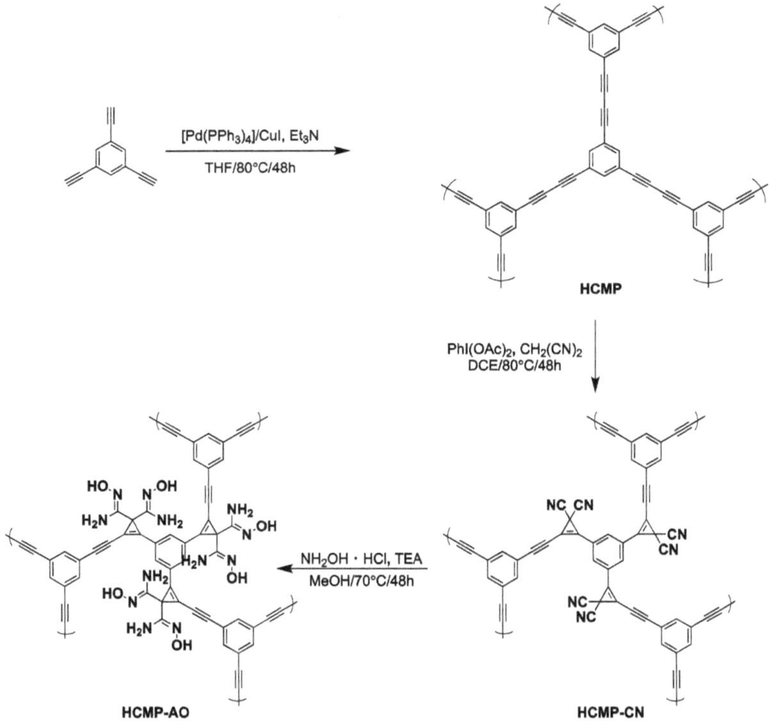

Figure 2.3 Synthesis of HCMP-1 functionalized with bis-amidoxime groups.

The original CMP was designed by condensing a single monomer, 1,3,5-trietbhynylbenzene, which was treated with malononitrile to give a bis-CN group-functionalized CMP network, HCMP-CN. To improve the uranium binding energy and adsorption ability, the

-CN group was converted to an amidoxime group, which has a high affinity for uranyl ions. Adsorption evaluation of uranium in aqueous solution using the obtained HCMP-AO showed that the ligands in CMP contributed to the original HCMP and HCMP-CN. This indicates that the bisamidoxime ligand in CMP has a unique coordination mode to the uranyl ion, suggesting that it is possible to enhance the affinity of the chelating group to a specific ion by synergistically coordinating the binding site by improving the coordination-binding interactions. These strategies are a promising way to design adsorbents with excellent trapping ability for various ionic pollutants, not limited to uranium adsorption.

2.3 Control Over Porous Structure

While many CMPs have three-dimensional micropores, covalent organic frameworks (COF) have basic units connected and extended to form uniform pore shapes. The ordered stacking of pores in the COF also forms aligned one-dimensional nano-channels, allowing molecules and ions to be smoothly transported along the one-dimensional nano-channels. On the other hand, because other CMPs are amorphous with irregular pore shapes, they usually have various pore shapes and wide pore distributions. Thus, theoretically, increasing surface and pore volume is beneficial to enhancing the absorption capacity. The porous structure of CMPs was primarily determined by the monomer structure, monomer ratio, reaction conditions, and post-modification approach.

2.3.1 Control by Tuning Monomer Structure and Ratio

The CMP skeleton consists of knots and linkers, the size of which greatly affects the porosity of the material. For example, under the condition that the knot size does not change, the pore structure of phenylethylene-based CMP can be adjusted by chaining the length of the linker [80]. Combining 1,3,5-trisubstituted phenyl derivatives with linkers of different lengths resulted in CMP-0, CMP-1, CMP-2, CMP-3, and CMP-5. As a result of systematic investigation, a remarkable length-dependent trend was observed in which the average opening increased with increasing linker length and the BET specific surface area gradually decreased to 1018–512 m^2/g

and the pore volume from 0.38 to 0.16 m^2/g. There was a similar relationship between phosphine oxide and 1,3,5-triphenylbenzene-based CMP, and an even shorter linker was shown to be conductive in the formation of microporous structures [81–84].

The shape of the connecting unit also plays an important role in the development of the porous structure. Figure 2.4 provides a series of CMPs cored with triphenylamine synthesized by Buchwald-Hartwig cross-coupling with pyridine blocks with different substituents and BET surface areas of 63.33 and 30 m^2/g when tris (4-bromophenyl) amine was copolymerized [85]. This result is consistent with the previously reported Yamamoto reaction of 2,2',7,7'-tetrabromo-9, 9'-spirofluorene ether with dibromobenzene, in which the substituent was changed [86]. Therefore, it is considered that the p-substituted units are advantageous for the formation of porous frameworks, and those with o or m substitutions are not introduced into the formation of porous networks due to the steric hindrance effect during polymerization.

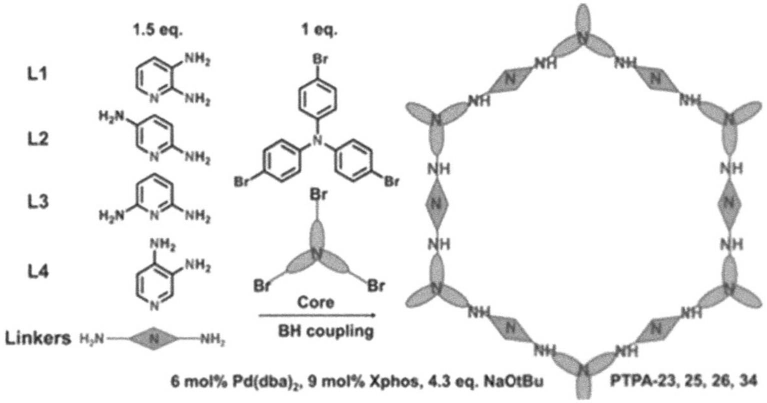

Figure 2.4 Synthetic route of PTPAs. Adapted from [85].

The ratio of knots to linkers greatly affects the porous structure of CMP. In the Yamamoto reaction with 2,2 '7,7'-tetrabromo-9, 9'-speirofluorene as the knot and 1,4-dibromobenzene as the linker, when the ratio of knot to linker was increased from 1: 0 to 1: 1, 1: 2, 1: 3, and 1: 4, the BET surface area, pore volume, and micropore volume decreased from 1275 m^2/g to 156 m^2/g, from 0. 997 to 0.159 cm^3/g, and from 0.749 to 0.097 cm^3/g, respectively [86]. In addition, the effect of the knot-to-linker ratio on the CMP porosity was also observed in the Sonogashira-Hagiwara cross-linking reaction of 1,3,5-triethylnylbenzene to 1,4-diethylbenzene [87], indicating that the surface area and total pore volume of CMP increase and the micropore volume decreases as the number of mesopores increases. These results indicate that the molar ratio of monomers usually determines the degree of polymerization of linear conjugated polymers and thus the porosity of CMPs.

2.3.2 Control by Tuning Reaction Conditions

The results show that the reaction conditions can greatly promote the development of porous structures in the materials. In CMPs, prepared by the Sonogashira-Hagihara reaction of 1,3,5-triethynylbenzene with different dibromo-substituted arene units, it was observed that the polarity of the solvent had a significant effect on porosity [88]. In the copolymerization of 1,3,5-triethynylbenzene and 1,4-dibromo-2, 5-difluorobenzene, the surface area of CMP tended to be 1260 to 869, 690, and 639 m^2/g in the corresponding solvents of DMF, dioxane, toluene, and THF, and the volume of micropores tended to be 0.48 to 0.39, 0.27, and 0.25 cm^3/g, respectively. On the other hand, the effect of the solvent was found to construct different degrees of porosity depending on the combination of specific monomers. When the reaction solvent was changed from toluene to 1,4-dioxane, the specific surface area of nitrogen-containing CMP (NCMP), which was synthesized by the palladium-catalyzed cross-condensation of 1,3,5-triethynylbenzene and bis (4-boromophenyl) amine, decreased from 945 to 593 m^2/g [89]. It was concluded that common solvents were not optimal and efforts to investigate this parameter were needed when developing new materials.

The catalyst affected the polymerization degree of CMPs, resulting in a series of networks with diverse pore sizes and surface areas. Taking the ion thermal synthesis of triazinyl polyporous polymer, as an example, the triazinyl polymer was synthesized from (1,1',4',1"-terphenyl)-4,4"-dicarbonitrile with 10- or 20-equivalent $ZnCl_2$ catalyst at 400 °C [90]. The BET surface areas were 495 and 1900 m^2/g, and the pore volumes were 0.97 and 2.25 cm^3/g for the reaction of 10- and 20-equivalent catalysts, indicating that the growth of catalyst dosage effectively improved the BET surface area and pore volume. Meanwhile, the 10-equivalent $ZnCl_2$ catalyst synthesis network was dominated by micropores while the 20-equivalent formed a large aperture mesoporous network.

The porosity of CMP also changed with reaction temperature. The BET surface area and pore volume increased from 920 m^2/g and 0.47 cm^3/g to 1600 m^2/g and 1.00 cm^3/g, 1750 m^2/g and 1.58 cm^3/g, 400 m^2/g, and 500 cm^3/g, respectively, with reference to the triazine network prepared by 1,4 disocyanobenzene at temperatures of 400, 500, 600, and 700 °C [90]. The micropore distribution in the sample synthesized at 400 °C was relatively narrow (1~2 nm) and that in the sample synthesized at 500 °C was relatively broad (1~6 nm).

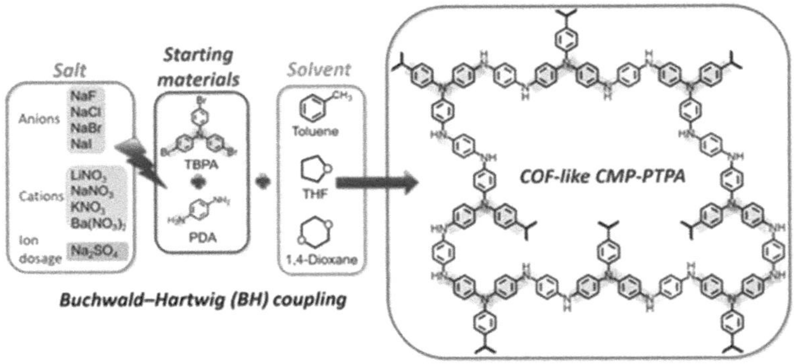

Figure 2.5 Synthetic route for the formation of salt-tunable PIPA networks. Adapted from [91].

Recent studies suggest that the porosity of NCMPs synthesized by Buchwald-Hartwig cross-couplings can be adjusted by the addition of salt, as shown in Figure 2.5 [91]. By the adoption of NaX (X = Cl, Br, I), the BET surface area of poly (triphenylamine) (PTPA) was improved by almost 20 times, from 58 m^2/g to 1134 m^2/g, and gradually increased as the ionic radius of the anion decreased. The addition of NaX also greatly improved the total pore volume, micropore volume, and ultra-micropore volume. Furthermore, the effect of the ionic radius of the salt was systematically investigated in $M_x (NO_3)_y$ (M = Li, Na, K, Ba). The BET surface area of $LuNO_3$-coated PTPA was 1152 m^2/g and increased with increasing the cation radius of the salt. From this, it is concluded that the introduction of salt affects the solubility parameters controlling the polymerization in all aspects, and the better the compatibility of the solvent with the polymer, the higher the degree of polymerization, the more pores, and the larger the specific surface area.

2.3.3 Control by Post-Modification Approach

The porosity of CMP can also be controlled by post-modification. CMP-1 was obtained by reacting 1,3,5-triacetylenebenzene with 1,4-dibromobenzene and then modified by varying amounts of thioethanol (C_2SH) or 6-mercaptohexanol (C_6SH) [92]. When reacted with 0.5, 1.0, and 1.5 equivalents of C_2SH per triple bond, the BET surface area gradually decreased from 1200 to 907 m^2/g and the pore volume from 0.465 to 0.352 cm^3/g. In the case of C_2SH, BET surface area and pore volume showed a similar decreasing trend. Porosity analysis showed that the microporosity could be adjusted to a certain range by varying the amount of modified mercaptan. In addition, thiadiazolyl CMP (BBT) also showed post-modification effects on BET surface area, resulting in BBT-SC_2NH_2 by thiolyne reaction of BBT with cysteamine (SC_2NH_2), increasing BET surface area from 110 to 148 m^2/g [93].

2.4 Perspectives and Challenges

This chapter introduces the main molecular design principles, synthesis, and structural studies of CMP. Among these, the monomer structure and ratio, reaction solvent, and post-modification were optimized to control the porous properties. From a theoretical point of view, the main strategies that can be implemented for further study of CMP design and pollutant removal include the following four points:

1. Although CMP has been extensively developed, its application to heavy metal removal is still limited due to its high hydrophobicity and difficulty in dispersing into water. To overcome this drawback, hydrophilic functional groups (-SO_3H, -OH, -COOH, -NH_2, etc.) can be introduced at the time of material design to improve the dispersibility to improve the adsorption performance of heavy metals. In addition, the introduction of polar groups mentioned above is effective in

improving the adsorption of gaseous pollutants by overcoming the relatively weak interaction with CMP.

2. The quest for controllable synthesis and mass production. In addition to being low-cost, a controllable synthesis ensures an engineering process. First, a soft template method has been established that can realize CMP with low cost, simple process, excellent reproducibility, regular structure, and unique shape, and has a wide range of possibilities. In general, CMP exists as an insoluble solid powder and has processing problems. To create a separation membrane, it is necessary to design a CMP that is easy to dissolve. Eventually, the invention of adsorptive membrane materials is expected to enable large-scale industrial production.
3. In theory, the essential properties of CMP determine the optimality for realizing the synergistic effect of physical and chemical adsorption by simultaneously adjusting pore size and pore chemistry and can address the challenge of simultaneous removal of multiple contaminants. Furthermore, the adoption of temperature-sensitive or light-sensitive CMP organogels seems promising for optimizing the balance between physical and chemical adsorption. Furthermore, we propose that the surface molecules of CMP shift from physical adsorption by weak van der Waals forces to strongly bound chemisorption states by overcoming the energy barrier, providing a new idea for the adsorption principle.
4. The application of resource recovery and recycling will focus on establishing an entirely new wastewater treatment concept with resource and energy recovery. On the other hand, to realize the recovery of heavy metals effectively, it is necessary to enhance the adsorption selectivity. The development of a process to separate, remove, and recover heavy metals from industrial wastewater provided a milestone for future wastewater treatment. On the other hand, reconciling the discrepancy between the adsorption and desorption performance of adsorbents and developing advanced light-renewable or biodegradable adsorbents would also be a major advantage for large-scale industrial applications.

References

1. S. W. Yuan, B. Dorney, D. White, S. Kirklin, P. Zapol, L. P. Yu, and D. J. Liu, *Chem. Commun.* 2010, 46, 4547–4549.
2. P. Kuhn, M. Antonietti, and A. Thomas, *Angew. Chem., Int. Ed.* 2008, 47, 3450–3453.
3. J. X. Jiang, F. Su, H. Niu, C. D. Wood, N. L. Campbell, Y. Z. Khimyak, and A. I. Cooper, *Chem. Commun.* 2008, 486–488.
4. J. Schmidt, M. Werner, and A. Thomas, *Macromolecules* 2009, 42, 4426–4429.
5. J. X. Jiang, A. Trewin, D. J. Adams, and A. I. Cooper, *Chem. Sci.*, 2011, 2, 1777–1781.
6. Y. H. Xu, L. Chen, Z. Q. Guo, A. Nagai, and D. Jiang, *J. Am. Chem. Soc.* 2011, 133, 17622–17625.
7. J. X. Jiang, F. B. Su, A. Trewin, C. D. Wood, N. L. Campbell, H. J. Niu, C. Dickinson, A. Y. Ganin, M. J. Rosseinsky, Y. Z. Khimyak, and A. I. Cooper, *Angew. Chem., Int. Ed.* 2007, 46, 8574–8578.
8. J. X. Jiang, F. B. Su, A. Trewin, C. D. Wood, H. J. Niu, J. T. A. Jones, Y. Z. Khimyak, and A. I. Cooper, *J. Am. Chem. Soc.* 2008, 130, 7710–7720.
9. G. Cheng, T. Hasell, A. Trewin, D. J. Adams, and A. I. Cooper, *Angew. Chem., Int. Ed.* 2012, 51, 12727–12731.
10. L. Chen, Y. Honsho, S. Seki, and D. Jiang, *J. Am. Chem. Soc.* 2010, 132, 6742–6748.
11. Q. Chen, M. Luo, T. Wang, J. X. Wang, D. Zhou, Y. Han, C. S. Zhang, C. G. Yan, and B. H. Han, *Macromolecules* 2011, 44, 5573–5577.
12. Y. Kou, Y. Xu, Z. Guo, and D. Jiang, *Angew. Chem., Int. Ed.*, 2011, 50, 8753–8757.
13. A. P. Côté, A. I. Benin, N. W. Ockwig, M. O'Keeffe, A. J. Matzger, and O. M. Yaghi, *Science* 2005, 310, 1166.
14. A. Kirchon, L. Feng, H. F. Drake, E. A. Joseph, and H.-C. Zhou, *Chem. Soc. Rev.*, 2018, 47, 8611–8638.
15. Z. Wang and S. M. Cohen, *J. Am. Chem. Soc.* 2007, 129, 12368–12369.
16. Y. Song, Q. Sun, B. Aguila, and S. Ma, *Adv. Sci.* 2018, 6, 1801410.
17. Y. Yusran, X. Guan, H. Li, Q. Fang, and S. Qiu, *Natl. Sci. Rev.* 2020, 7, 170–190.
18. Jiang, J.-X., Su, F., Trewin, A., Wood, C. D., Campbell, N. L., Niu, H., Dickinson, C., Ganin, A. Y., Rosseinsky, M. J., Khimyak, Y. Z., et al. Conjugated Microporous Poly(Aryleneethynylene) Networks. *Angew. Chem., Int. Ed.* 2007, 46, 8574–8578.

19. Ben, T., Ren, H., Ma, S., Cao, D., Lan, J., Jing, X., Wang, W., Xu, J., Deng, F., Simmons, J. M., et al. Targeted Synthesis of a Porous Aromatic Framework with High Stability and Exceptionally High Surface Area. *Angew. Chem., Int. Ed.* 2009, 48, 9457–9460.

20. Weber, J. and Thomas, A. Toward Stable Interfaces in Conjugated Polymers: Microporous Poly(P-Phenylene) and Poly-(Phenyleneethynylene) Based on a Spirobifluorene Building Block. *J. Am. Chem. Soc.* 2008, 130, 6334–6335.

21. Jiang, J.-X., Laybourn, A., Clowes, R., Khimyak, Y. Z., Bacsa, J., Higgins, S. J., Adams, D. J., and Cooper, A. I. High Surface Area Contorted Conjugated Microporous Polymers Based on Spiro-Bipropylenedioxythiophene. *Macromolecules* 2010, 43, 7577–7582.

22. Chen, Q., Wang, J.-X., Wang, Q., Bian, N., Li, Z.-H., Yan, C.-G., and Han, B.-H. Spiro(Fluorene-9,9′-Xanthene)-Based Porous Organic Polymers: Preparation, Porosity, and Exceptional Hydrogen Uptake at Low Pressure. *Macromolecules* 2011, 44, 7987–7993.

23. Jiang, J.-X., Laybourn, A., Clowes, R., Khimyak, Y. Z., Bacsa, J., Higgins, S. J., Adams, D. J., and Cooper, A. I. High Surface Area Contorted Conjugated Microporous Polymers Based on Spiro-Bipropylenedioxythiophene. *Macromolecules* 2010, 43, 7577–7582.

24. Chen, Q., Wang, J.-X., Wang, Q., Bian, N., Li, Z.-H., Yan, C.-G., and Han, B.-H. Spiro(Fluorene-9,9′-Xanthene)-Based Porous Organic Polymers: Preparation, Porosity, and Exceptional Hydrogen Uptake at Low Pressure. *Macromolecules* 2011, 44, 7987–7993.

25. Doucet, H. and Hierso, J.-C. Palladium-Based Catalytic Systems for the Synthesis of Conjugated Enynes by Sonogashira Reactions and Related Alkynylations. *Angew. Chem., Int. Ed.* 2007, 46, 834–871.

26. Trunk, M., Herrmann, A., Bildirir, H., Yassin, A., Schmidt, J., and Thomas, A. Copper-Free Sonogashira Coupling for High-Surface-Area Conjugated Microporous Poly(Aryleneethynylene) Networks. *Chem. Eur. J.* 2016, 22, 7179–7183.

27. Jiang, J.-X., Su, F., Niu, H., Wood, C. D., Campbell, N. L., Khimyak, Y. Z., and Cooper, A. I. Conjugated Microporous Poly-(Phenylene Butadiynylene). *Chem. Commun.* 2008, 486–488.

28. Thorand, S. and Krause, N. Improved Procedures for the Palladium-Catalyzed Coupling of Terminal Alkynes with Aryl Bromides (Sonogashira Coupling). *J. Org. Chem.* 1998, 63, 8551–8553.

29. Miyaura, N., Yamada, K., and Suzuki, A. A New Stereospecific Cross-Coupling by the Palladium-Catalyzed Reaction of 1-Alkenyl-boranes

with 1-Alkenyl or 1-Alkynyl Halides. *Tetrahedron Lett.* 1979, 20, 3437–3440.

30. Miyaura, N. and Suzuki, A. Stereoselective Synthesis of Arylated (E)-Alkenes by the Reaction of Alk-1-Enylboranes with Aryl Halides in the Presence of Palladium Catalyst. *J. Chem. Soc., Chem. Commun.* 1979, 866–867.

31. Chen, L., Honsho, Y., Seki, S., and Jiang, D. Light-Harvesting Conjugated Microporous Polymers: Rapid and Highly Efficient Flow of Light Energy with a Porous Polyphenylene Framework as Antenna. *J. Am. Chem. Soc.* 2010, 132, 6742–6748.

32. Liu, Q., Tang, Z., Wu, M., and Zhou, Z. Design, Preparation and Application of Conjugated Microporous Polymers. *Polym. Int.* 2014, 63, 381–392.

33. Schmidt, J., Werner, M., and Thomas, A. Conjugated Microporous Polymer Networks Via Yamamoto Polymerization. *Macromolecules* 2009, 42, 4426–4429.

34. Jiang, J.-X., Trewin, A., Adams, D. J., and Cooper, A. I. Band Gap Engineering in Fluorescent Conjugated Microporous Polymers. *Chem. Sci.* 2011, 2, 1777–1781.

35. Schmidt, J., Werner, M., and Thomas, A. Conjugated Microporous Polymer Networks Via Yamamoto Polymerization. *Macromolecules* 2009, 42, 4426–4429.

36. Ben, T., Ren, H., Ma, S., Cao, D., Lan, J., Jing, X., Wang, W., Xu, J., Deng, F., Simmons, J. M., et al. Targeted Synthesis of a Porous Aromatic Framework with High Stability and Exceptionally High Surface Area. *Angew. Chem., Int. Ed.* 2009, 48, 9457–9460.

37. Sun, L., Zou, Y., Liang, Z., Yu, J., and Xu, R. A One-Pot Synthetic Strategy Via Tandem Suzuki-Heck Reactions for the Construction of Luminescent Microporous Organic Polymers. *Polym. Chem.* 2014, 5, 471–478.

38. Yuan, S., Dorney, B., White, D., Kirklin, S., Zapol, P., Yu, L., and Liu, D.-J. Microporous Polyphenylenes with Tunable Pore Size for Hydrogen Storage. *Chem. Commun.* 2010, 46, 4547–4549.

39. Kuhn, P., Antonietti, M., and Thomas, A. Porous, Covalent Triazine-Based Frameworks Prepared by Ionothermal Synthesis. *Angew. Chem., Int. Ed.* 2008, 47, 3450–3453.

40. Katekomol, P., Roeser, J., Bojdys, M., Weber, J., and Thomas, A. Covalent Triazine Frameworks Prepared from 1,3,5-Tricyanobenzene. *Chem. Mater.* 2013, 25, 1542–1548.

41. Hao, L., Luo, B., Li, X., Jin, M., Fang, Y., Tang, Z., Jia, Y., Liang, M., Thomas, A., Yang, J., et al. Terephthalonitrile-Derived Nitrogen-Rich Networks for High Performance Supercapacitors. *Energy Environ. Sci.* 2012, 5, 9747–9751.

42. Ren, S., Bojdys, M. J., Dawson, R., Laybourn, A., Khimyak, Y. Z., Adams, D. J., and Cooper, A. I. Porous, Fluorescent, Covalent Triazine-Based Frameworks Via Room-Temperature and Microwave-Assisted Synthesis. *Adv. Mater.* 2012, 24, 2357–2361.

43. Wang, K., Yang, L.-M., Wang, X., Guo, L., Cheng, G., Zhang, C., Jin, S., Tan, B., and Cooper, A. Covalent Triazine Frameworks Via a Low-Temperature Polycondensation Approach. *Angew. Chem., Int. Ed.* 2017, 56, 14149–14153.

44. Yu, S.-Y., Mahmood, J., Noh, H.-J., Seo, J.-M., Jung, S.-M., Shin, S.-H., Im, Y.-K., Jeon, I.-Y., and Baek, J.-B. Direct Synthesis of a Covalent Triazine-Based Framework from Aromatic Amides. *Angew. Chem., Int. Ed.* 2018, 57, 8438–8442.

45. Buyukcakir, O., Yuksel, R., Jiang, Y., Lee, S. H., Seong, W. K., Chen, X., and Ruoff, R. S. Synthesis of Porous Covalent Quinazoline Networks (CQNs) and Their Gas Sorption Properties. *Angew. Chem., Int. Ed.* 2019, 58, 872–876.

46. Stille, J. K. and Mainen, E. Ladder Polyquinoxalines. *J. Polym. Sci., Part B: Polym. Lett.* 1966, 4, 39–41.

47. Stille, J. K. and Mainen, E. L. Thermally Stable Ladder Polyquinoxalines. *Macromolecules* 1968, 1, 36–42.

48. Kou, Y., Xu, Y., Guo, Z., and Jiang, D. Supercapacitive Energy Storage and Electric Power Supply Using an Aza-Fused Π-Conjugated Microporous Framework. *Angew. Chem., Int. Ed.* 2011, 50, 8753–8757.

49. Marco, A. B., Cortizo-Lacalle, D., Perez-Miqueo, I., Valenti, G., Boni, A., Plas, J., Strutyński, K., DeFeyter, S., Paolucci, F., Montes, M., et al. Twisted Aromatic Frameworks: Readily Exfoliable and Solution-Processable Two-Dimensional Conjugated Microporous Polymers. *Angew. Chem., Int. Ed.* 2017, 56, 6946–6951.

50. Ding, S.-Y. and Wang, W. Covalent Organic Frameworks (COFs): From Design to Applications. *Chem. Soc. Rev.* 2013, 42, 548–568.

51. Pandey, P., Katsoulidis, A. P., Eryazici, I., Wu, Y., Kanatzidis, M. G., and Nguyen, S. T. Imine-Linked Microporous Polymer Organic Frameworks. *Chem. Mater.* 2010, 22, 4974–4979.

52. Rabbani, M. G., Sekizkardes, A. K., El-Kadri, O. M., Kaafarani, B. R., and El-Kaderi, H. M. Pyrene-Directed Growth of Nanoporous Benzimidazole-

Linked Nanofibers and Their Application to Selective CO_2 Capture and Separation. *J. Mater. Chem.* 2012, 22, 25409–25417.

53. Xu, C. and Hedin, N. Synthesis of Microporous Organic Polymers with High CO_2-over-N_2 Selectivity and CO_2 Adsorption. *J. Mater. Chem. A* 2013, 1, 3406–3414.

54. Rabbani, M. G. and El-Kaderi, H. M. Template-Free Synthesis of a Highly Porous Benzimidazole-Linked Polymer for CO_2 Capture and H_2 Storage. *Chem. Mater.* 2011, 23, 1650–1653.

55. Biswal, B. P., Becker, D., Chandrasekhar, N., Seenath, J. S., Paasch, S., Machill, S., Hennersdorf, F., Brunner, E., Weigand, J. J., Berger, R., et al. Exploration of Thiazolo[5,4-D]Thiazole Linkages in Conjugated Porous Organic Polymers for Chemoselective Molecular Sieving. *Chem. Eur. J.* 2018, 24, 10868–10875.

56. Zhu, X., Tian, C., Jin, T., Wang, J., Mahurin, S. M., Mei, W., Xiong, Y., Hu, J., Feng, X., Liu, H., et al. Thiazolothiazole-Linked Porous Organic Polymers. *Chem. Commun.* 2014, 50, 15055–15058.

57. Pennella, F., Banks, R. L., and Bailey, G. C. Disproportionation of Alkynes. *Chem. Commun.* 1968, 1548–1549.

58. Lu, G., Yang, H., Zhu, Y., Huggins, T., Ren, Z. J., Liu, Z., and Zhang, W. Synthesis of a Conjugated Porous Co(II) Porphyrinylene-Ethynylene Framework through Alkyne Metathesis and Its Catalytic Activity Study. *J. Mater. Chem. A* 2015, 3, 4954–4959.

59. Jiang, J.-X., Su, F., Niu, H., Wood, C. D., Campbell, N. L., Khimyak, Y. Z., and Cooper, A. I. Conjugated Microporous Poly-(Phenylene Butadiynylene). *Chem. Commun.* 2008, 486–488.

60. Li, B., Guan, Z., Yang, X., Wang, W.D., Wang, W., Hussain, I., Song, K., Tan, B., and Li, T. Multifunctional Microporous Organic Polymers. *J. Mater. Chem. A* 2014, 2, 11930–11939.

61. Zhang, Y., A, S., Zou, Y., Luo, X., Li, Z., Xia, H., Liu, X., and Mu, Y. Gas Uptake, Molecular Sensing and Organocatalytic Performances of a Multifunctional Carbazole-Based Conjugated Microporous Polymer. *J. Mater. Chem. A* 2014, 2, 13422–13430.

62. Vingiello, F. A., Yanez, J., and Campbell, J. A. New Approach to the Synthesis of Dibenzo[a,l] Pyrenes. *J. Org. Chem.* 1971, 36, 2053–2056.

63. Liao, Y., Wang, H., Zhu, M., and Thomas, A. Efficient Supercapacitor Energy Storage Using Conjugated Microporous Polymer Networks Synthesized from Buchwald–Hartwig Coupling. *Adv. Mater.* 2018, 30, 1705710.

64. Liao, Y., Weber, J., and Faul, C. F. J. Conjugated Microporous Polytriphenylamine Networks. *Chem. Commun.* 2014, 50, 8002–8005.

65. Gu, C., Chen, Y., Zhang, Z., Xue, S., Sun, S., Zhang, K., Zhong, C., Zhang, H., Pan, Y., Lv, Y., et al. Electrochemical Route to Fabricate Film-Like Conjugated Microporous Polymers and Application for Organic Electronics. *Adv. Mater.* 2013, 25, 3443–3448.

66. Gu, C., Huang, N., Gao, J., Xu, F., Xu, Y., and Jiang, D. Controlled Synthesis of Conjugated Microporous Polymer Films: Versatile Platforms for Highly Sensitive and Label-Free Chemo- and Biosensing. *Angew. Chem., Int. Ed.* 2014, 53, 4850–4855.

67. Zhang, H., Zhang, Y., Gu, C., and Ma, Y. Electropolymerized Conjugated Microporous Poly(Zinc-Porphyrin) Films as Potential Electrode Materials in Supercapacitors. *Adv. Energy Mater.* 2015, 5, 1402175.

68. Gu, C., Huang, N., Chen, Y., Qin, L., Xu, H., Zhang, S., Li, F., Ma, Y., and Jiang, D. Π-Conjugated Microporous Polymer Films: Designed Synthesis, Conducting Properties, and Photoenergy Conversions. *Angew. Chem., Int. Ed.* 2015, 54, 13594–13598.

69. Germain, J., Frechet, J. M. J., and Svec, F. Hypercrosslinked Polyanilines with Nanoporous Structure and High Surface Area: Potential Adsorbents for Hydrogen Storage. *J. Mater. Chem.* 2007, 17, 4989–4997.

70. Germain, J., Frechet, J. M. J., and Svec, F. Nanoporous, Hypercrosslinked Polypyrroles: Effect of Crosslinking Moiety on Pore Size and Selective Gas Adsorption. *Chem. Commun.* 2009, 1526–1528.

71. B. F. Hoskins and R. Robson, *J. Am. Chem. Soc.* 1990, 112, 1546–1554.

72. Z. Wang and S. M. Cohen, *J. Am. Chem. Soc.* 2007, 129, 12368–12369.

73. Y. Y. Cui, C. X. Yang, and X. P. Yan, *ACS Appl. Mater. Interfaces* 2020, 12, 4954–4961.

74. Y. Zhang and S. N. Riduan, *Chem. Soc. Rev.* 2012, 41, 2083–2094.

75. J. L. Segura, S. Royuela, and M. M. Ramos, *Chem. Soc. Rev.* 2019, 48, 3903–3945.

76. L. Cheng, Y. Yang, and D. Jiang, *J. Am. Chem. Soc.* 2010, 132, 9138–9143.

77. J. X. Jiang, C. Wang, A. Laybourn, T. Hasell, B. Clowes, and A. I. Cooper. *Angew. Chem., Int. Ed.* 2011, 50, 1072–1075.

78. J. X. Jiang, F. B. Su, A. Trewin, C. D. Wood, N. L. Campbell, H. J. Niu, C. Dickinson, A. Y. Ganin, M. J. Rosseinsky, Y. Z. Khimyak, and A. I. Cooper. *Angew. Chem., Int. Ed.* 2007, 46, 8574–8578.

79. L. Zhang, N. Pu, B. Yu, G. Ye, J. Chen, S. Xu, and S. Ma. *ACS Appl. Mater. Interfaces* 2020, 12, 3688–3696.

80. Jiang, J., Su, F., Trewin, A., Niu, H., Jones, J., Khimyak, Y., and Cooper, A. Synthetic Control of the Pore Dimension and Surface Area in Conjugated Microporous Polymer and Copolymer Networks. *J. Am. Chem. Soc.* 2008, 130, 7710–7720.

81. Jiang, J., Trewin, A., Adams, D., and Cooper, A. Band Gap Engineering in Fluorescent Conjugated Microporous Polymers. *Chem. Sci.* 2011, 2, 1777–1781.

82. Wang, Z., Yang, X., Yang, T., Zhao, Y., Wang, F., Chen, Y., Zeng, J., Yan, C., Huang, F., and Jiang, J. Dibenzothiophene Dioxide-Based Conjugated Microporous Polymers for Visible-Light-Driven Hydrogen Production. *ACS Catal.* 2018, 9, 8590–8596.

83. Qiao, S., Huang, W., Du, Z., Chen, X., Shiehc, F., and Yang, R. Phosphine Oxide-Based Conjugated Microporous Polymers with Excellent CO_2 Capture Properties. *New J. Chem.* 2015, 39, 136–141.

84. Qian, X., Wang, B., Zhu, Z., Sun, H., Ren, F., Mu, P., Ma, C., Liang, W., and Li, A. Novel N-Rich Porous Organic Polymers with Extremely High Uptake for Capture and Reversible Storage of Volatile Iodine. *J. Hazard. Mater.* 2017, 5, 244–232.

85. Li, H., Lyu, W., and Liao, Y. Engineering Redox Activity in Conjugated Microporous Polytriphenylamine Networks Using Pyridyl Building Blocks Toward Efficient Supercapacitors. *Macromol. Rapid Commun.* 2019, 40, 1900455.

86. Schmidt, J., Werner, M., and Thomas, A. Conjugated Microporous Polymer Networks Via Yamamoto Polymerization. *Macromolecules* 2009, 42, 4426–4429.

87. Tan, D., Fan, W., Xiong, W., Sun, H., Li, A., Deng, W., and Meng, C. Study on Adsorption Performance of Conjugated Microporous Polymers for Hydrogen and Organic Solvents: The Role of Pore Volume. *Eur. Polym. J.* 2012, 48, 705–711.

88. Dawson, R., Laybourn, A., Khimyak, Y., Adams, D., and Cooper, A. High Surface Area Conjugated Microporous Polymers: The Importance of Reaction Solvent Choice. *Macromolecules* 2010, 43, 8524–8530.

89. Mu, P., Sun, H., Zhu, Z., Liang, W., Liu, J., and Li, A. Synthesis and Properties of Nitrogen-Containing Conjugated Microporous Polymers. *Macromol. Mater. Eng.* 2016, 301, 451–456.

90. Kuhn, P., Thomas, A., and Antonietti, M. Toward Tailorable Porous Organic Polymer Networks: A High-Temperature Dynamic Polymerization Scheme Based on Aromatic Nitriles. *Macromolecules* 2009, 42, 319–326.

91. Chen, J., Yan, W., Townsend, E., Feng, J., Pan, J., Del Angel Hernandez, V., and Faul, J. Tuneable Surface Area, Porosity and Function in Conjugated Microporous Polymers. *Angew. Chem. Int. Ed.* 2019, 58, 11715–11719.

92. Kiskan, B. and Weber, J. Versatile Post-Modification of Conjugated Microporous Polymers Using Thiolyne Chemistry. *ACS Macro Lett.* 2012, 1, 37–40.

93. Wang, Q., Li, R., Ouyang, X., and Wang, G. A Novel Indole-Based Conjugated Microporous Polymer for Highly Active Removal of Heavy Metals from Aqueous Solution Via Double Cation–Π Interactions. *RSC Adv.* 2019, 9, 40531–40535.

94. Y. Xu, S. Jin, A. Nagai, and D. Jiang, *Chem. Soc. Rev.*, 2013, 42, 8012–8031.

Chapter 3

Surface Area and the Theory of Porous Materials

Because porous polymers are both soluble and amorphous, characterization of these materials is very difficult. One of the most common characterization methods applied to all porous materials, not just these materials, is gas sorption. When gas is adsorbed onto a porous material, the pore size, volume, and surface area of the material are very visible. This is why gas sorption is an important characterization technique, from crystalline materials such as MOF and COF to amorphous porous materials such as CMP, HCP, and CTF.

3.1 Introduction

Microporous polymers (MOPs) containing CMPs are a relatively new class of polymers that can be applied as adsorbents and polymer films in gas separation processes. These polymers are composed of pure organic building blocks with covalent bonds. The polymer can be synthesized by catalysis with or without a metal center. In recent years, these polymers have been extensively studied as adsorbents or membrane materials to separate CO_2 from flue gas. CO_2 separation from flue gas by adsorption and membrane methods may provide a cost-effective alternative to the methods currently used. This

Conjugated Microporous Polymers
Atsushi Nagai

ISBN 978-981-5129-66-3 (Hardcover), 978-1-003-63652-6 (eBook)
www.jennystanford.com

chapter describes the gas sorption theory of porous materials with both crystalline and amorphous structures.

3.2 Gas Sorption Theory

Gases (sorbate) can interact with the surface of a solid (sorbent) in a variety of ways. These interactions can be strong (chemisorption) with the formation of chemical bonds or weak (physisorption) with only van der Waals interactions. When we model the size and volume of pores, we are mostly concerned with the physical adsorption of ideal gases onto porous materials at low temperatures. In most cases, this is 77 K of nitrogen gas.

There are many different models that can be applied to the adsorption of a sorbate onto a sorbent and the first of them to be developed was the Langmuir model.

3.2.1 Langmuir Model

In 1916, Langmuir set out his adsorption isotherm, which related the adsorption of a gas molecule onto a solid surface as a function of pressure [1]. He proposed that a dynamic equilibrium exists between adsorbed and free gas molecules. This can be expressed as:

$$A(g) + B(s) \underset{\text{Desorption}}{\overset{\substack{\text{Adsorption} \\ K_{\text{ads}}}}{\rightleftharpoons}} AB$$

where $A\ (g)$ is a free molecule, B is an unoccupied surface area, and AB is an adsorbed gas molecule occupying a surface site. K_{ads} and K_{des} represent the respective rates of the adsorption and desorption processes.

This kinetic theory can be used to determine the rate of two processes and equilibrium, where the two rates are equal the overall rate constant in equilibrium is equivalent to:

$$K = \frac{K_{\text{ads}}}{K_{\text{des}}} = \frac{[AB]}{[A][B]}$$

At equilibrium what this means is that the rate of desorption from occupied sites is equal to that of adsorption to the occupied sites at a specific pressure. This can be expressed as:

$$K_{\mathrm{des}}\theta = K_{\mathrm{ads}}P(1-\theta)$$

where θ is the fractional surface coverage.

This equation can be rearranged to give θ and doing this results in:

$$\theta = \frac{K_{\mathrm{ads}}P}{K_{\mathrm{des}} + K_{\mathrm{ads}}P}$$

Remembering that $K = K_{\mathrm{abs}}/K_{\mathrm{des}}$ we get the following:

$$\theta = \frac{KP}{1+KP}$$

This is well-known as the Langmuir adsorption equation. Although acceptable when expressed in this form, the gas sorption theory seems appropriate to express this in terms of the volume of gas adsorbed by a sorbent. This gives the following:

$$\theta = \frac{V_{\mathrm{ads}}}{V_{\mathrm{mono}}}$$

where V_{abs} is the volume of gas adsorbed and V_{mono} is the amount of gas adsorbed corresponding to monolayer coverage.

This equation can then be re-written and arranged to form a linear equation, which yields the following:

$$\frac{P}{V} = \frac{P}{V_{\mathrm{mono}}} + \frac{1}{KV_{\mathrm{mono}}}$$

If a graph of P/V vs V was plotted, a straight line where the slope would be equal to $1/V_{\mathrm{mono}}$ and the intercept $1/KV_{\mathrm{mono}}$.

Although the Langmuir model is a useful model for gas sorption, it is limited in that a number of assumptions are made when applying this model. Namely, these are:

1. Dynamic equilibrium exits between adsorbed and free gas molecules.
2. The surface of the sorbent is uniform and the enthalpy of adsorption does not vary with coverage.
3. Adsorption is limited to only monolayer coverage – after this no further adsorption is witnessed.
4. There are no significant interactions between adsorbed molecules.

These assumptions limit the effectiveness of the Langmuir model when trying to determine the surface area or pore size of a solid. For example, the surface of most materials is not uniform, it is often heterogeneous, and assumption 2 is invalid. It is also known that the surface coating is not limited to a single layer so assumption 3 is invalid. This only really happens at low pressures. Similarly, assumption 4 is invalid because attractive forces exist between molecules of the same type. For these reasons, the Langmuir model for gas adsorption is valid only under conditions of low pressure, and above that, another gas adsorption model must be applied. A different gas adsorption model needs to be applied.

3.2.2 Brunauer-Emmett-Teller (BET) Theory

The Brunauer-Emmett-Teller (BET) theory, named after the scientist who developed it, was formulated on the basis that adsorption could extend beyond monolayer formation to more than one layer in thickness [2]. Under high-pressure and low-temperature conditions, a multilayer structure is formed because the thermal energy decreases and there are more gas molecules. Like the Langmuir adsorption model, the BET theory has the following assumptions:

1. Gas molecules adsorb onto a solid infinitely – one monolayer layer need not be complete before another begins.
2. Only considering interaction is that an adsorbed molecule acts as a single adsorption site for a molecule on the upper layer.
3. The molar heat of adsorption for the first layer is higher than that for the succeeding layers – the first layer is equal to the enthalpy of adsorption; whereas, the other layers are equal to the enthalpy of liquefaction.
4. Langmuir adsorption theory can be applied to each adsorbed layer.

The overall equation for the BET theory, which takes all these assumptions into account, is presented as:

$$\theta = \frac{c(P/P_0)}{(1-P/P_0)(1+((c-1)P/P_0))}$$

where P is the equilibrium gas pressure, P_0 is the saturated vapor pressure, P/P_0 is the relative pressure, and c is a constant relating to the energy of adsorption:

$$c = e^{(\Delta H_{\text{ads}} - \Delta H_{\text{liq}})/RT}$$

Once more considering the volume of gas adsorbed the equation can be expressed in linear form as:

$$\frac{(P/P_0)}{V(1-(P/P_0))} = \frac{1}{cV_{\text{m}}} + \frac{c-1}{cV_{\text{m}}}(P/P_0)$$

where V is the volume of adsorbed gas at pressure P and V_{m} is the volume of gas required to form a monolayer.

A plot of the left-hand side of the equation against P/P_0 gives a linear plot over a relative pressure range of between 0.05 and 0.35 when the BET equation is valid with:

$$\text{Slope} = \frac{c-1}{cV_{\text{m}}} \qquad \text{Intercept} = \frac{1}{cV_{\text{m}}}$$

The BET model considers multilayer adsorption on inhomogeneous surfaces and has been applied to gas adsorption on porous materials. In most cases, inert gases, usually boiling point (77 K) nitrogen, have been the adsorbent of choice. When N_2 is used for gas adsorption experiments on solid adsorbents, the surface area of the material can be calculated using the assumed molecular weight of N_2 molecules of 1.62×10^{-19} m^2. The following equation can be used to determine the specific surface area of BET per gram of sample:

$$SA_{\text{BET}} = \frac{PV_m}{RT} N\alpha A$$

where R is the ideal gas constant (8.31 J mol^{-1} K^{-1}), T is the temperature, N is Avogadro's number, and A is the area an N_2 molecule occupies (1.62×10^{-19} m^2).

3.2.3 Isotherm Shapes

In 1985, IUPAC divided the shapes of nitrogen gas sorption isotherms into six distinct types [3]. These are shown in Figure 3.1.

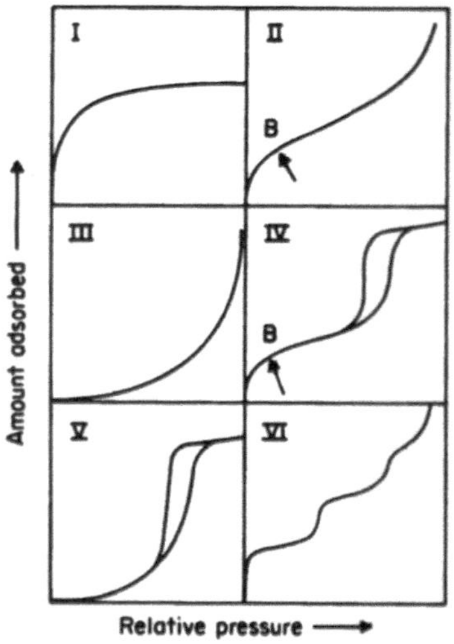

Figure 3.1 IUPAC isotherm classifications from 1985 [3].

These six types are distinguished by certain pore structures and are described as such. Type I isotherms are pure microporous materials with a few external surfaces, showing very large uptake at low pressures. In this case, the limit of uptake is governed by the volume of accessible micropores, not the internal surface area. Type II isotherms are given by nonporous or macroporous adsorbents and represent unlimited single-layer to multilayer adsorption. Point B represents the point where the coating of the monolayer is complete. Type III isotherms have a convex shape with respect to the P/P_0 axis and are very rare. The adsorbent-adsorbent cohesive force represents a material that is stronger than the adsorbent-adsorbent adhesive force. Type IV isotherms are given by mesopores and are characterized by hysteresis loops associated with capillary condensation occurring in such mesopores. It is initially similar to Type I and Type II isotherms because of single-layer to multi-layer adsorption. Despite the weak cohesive force between the adsorbent and the adsorbent, the adsorbent is adsorbed in several pores. Finally, Type VI represents stepwise multilayer adsorption onto a uniform non-porous surface. The step height represents the single-layer capacity of each adsorption layer.

For many years, these six types of isotherms were, and still are, the shapes expected when conducting gas sorption experiments.

However, over the past 40 years, as the field has continuously evolved, more and more groups have generated more and more isotherms for many different and diverse materials. This led to a review of the types of isotherms, and in 2015 IUPAC published an updated version of the original classification [4]. The proposed IUPAC classification of isotherm types is represented in Figure 3.2.

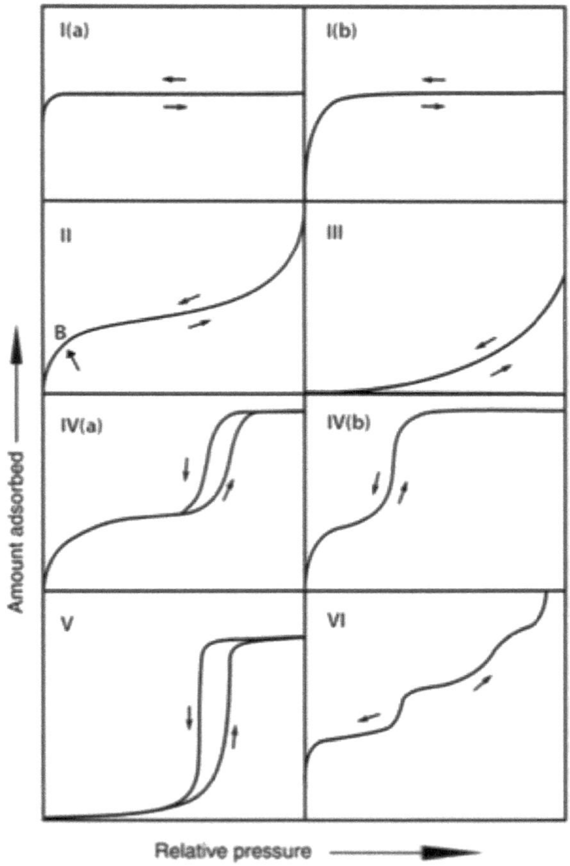

Figure 3.2 Revised IUPAC isotherm classifications from 2015 [4].

The type of isotherm is almost the same as in the 1985 classification, but the 2 main differences affect the Type I and Type IV isotherms. Type I isotherms have come to be divided into Type I (a), represented mainly by microporous materials with narrow pore sizes (width < 1 nm), and Type I (b), represented by microporous materials with wider pore size distributions and possibly mesopores (1–2.5 nm). A clear difference in shape is found in the low relative pressure region, where there is more uptake in the lower relative pressure region in Type I (a); whereas, in Type I (b), it extends to the relative pressure region where uptake is slightly larger. Type IV, like Type I, is divided into two categories: IV (a) and IV (b). The difference between the two isotherms is due to the size of the mesopore that causes capillary condensation, creating the classic hysteresis loop associated with this type. The isotherm of Type IV (a) has a remarkably large hysteresis loop, which is a result of the presence of mesopores larger than 4 nm in the sample. Adsorbents with smaller pore widths exhibit Type IV (b) isotherms in which hysteresis is less pronounced or absent at all because capillary condensation is very small.

3.2.4 Adsorption Hysteresis

The presence of hysteresis loops in the multilayer physical adsorption regions of isotherms is generally associated with capillary condensation [5]. This form of hysteresis occurs mainly in the mesopores of the material and is attributed to network effects or metastability of adsorption. For example, in open-ended pores, delayed condensation is a result of the metastability of the adsorption layer. This means that the adsorbed branches of the isotherm are not in thermodynamic equilibrium. Instead, when the pores are filled with liquid condensate, equilibrium is established during desorption, which is why hysteresis is observed.

It is shown that there is a direct correlation between the shape of the observed hysteresis and the textural properties of the adsorbent. It was classified by IUPAC in 1985 and updated in 2015. Figure 3.3 shows the six types of hysteresis loops observed in gas sorption studies.

H1 loops are typically found in materials with a narrow range of uniform pores but have also been reported in networks with pores with neck widths smaller than the cavity, such as ink bottles. This loop shows delayed condensation at the adduct. H2-type loops are the result of complex pore structures dominated by network effects and can be divided into two types. The H2 (a) loop has a characteristic steep desorption branch and is the result of evaporation due to hole blockage or cavitation. The H2 (b) loop is caused by the low steepness of the desorption branch and the pore blocking, but in this case, the pore neck width is very large. The H3 loop has two characteristic features: (i) the adsorption branch resembles a Type II isotherm, and (ii) the lower limit of the desorption branch is located in the P/P_0 range due to cavitation. The H4 loop is similar to the H3 loop, but the adsorption branches are closer to Type I, which is due to more micropores being filled. Finally, the H5 loop has a unique shape because of the presence of both open and partially blocked mesopores.

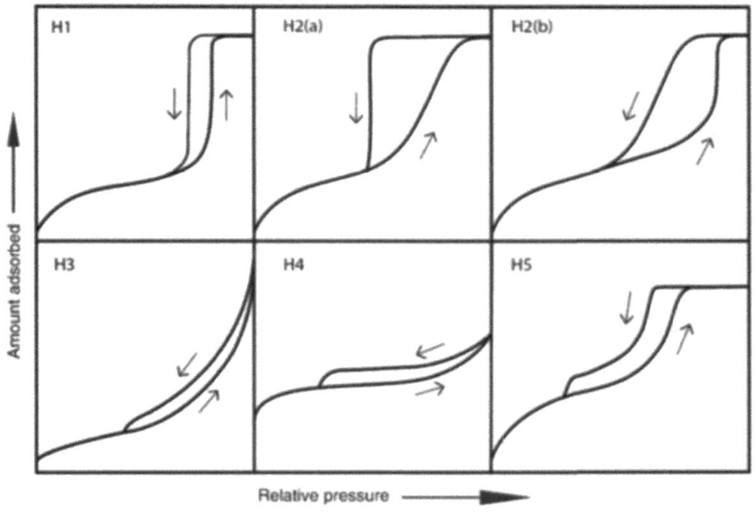

Figure 3.3 Observed hysteresis loops as classified by IUPAC.

Complementary reasoning for the occurrence of hysteresis is provided by network models that deal with so-called pore-blocking (percolation) effects and cavitation phenomena [6–8]. Pore-blocking is assumed to take place in pores with a wide body and a narrow neck. Although the wide body would empty at higher relative pressure, evaporation of its contents is kinetically obstructed by the narrow entrance, which itself is filled with adsorbate that requires lower P/P_0 to be emptied. Thus, the wide pore body remains filled until the neck is emptied. When that has been achieved, the entire contents of the body evaporate at once causing the isotherm to plummet almost vertically. The other phenomenon is called cavitation and is considered to occur in pores with necks of yet smaller diameters than is the case for pore-blocking. In that instance, the adsorbate inside the neck demands such low relative pressure that the contents of the wide body, which is very eager to evaporate at much higher relative pressure, cannot be held in check and start to evaporate irrespective of the neck still being filled. This happens at a relative pressure higher than required for emptying the neck and proceeds via cavitation, i.e., gas bubbles formed in the pore body diffuse through the neck that is still filled with liquid adsorbate. As a result, the pore body will be emptied; whereas, the neck is still filled. The neck then starts to release its adsorbate when the respective P/P_0 is reached.

However, other gases have similar applications in sorption analysis. In addition, techniques other than gas sorption have proven useful, and this section sheds light on some of them.

Sorption analysis of carbon dioxide, usually performed at 273/298 K, is particularly suitable for the characterization of materials containing very small pores, so-called ultrafine pores, less than 0.7 nm in diameter, that are inaccessible to nitrogen. It also has the advantage of requiring less of the machine because it can be measured at temperatures close to room temperature and in a more pronounced vacuum. In addition, because of the relatively high measurement temperature, the gas has high thermal and kinetic energy, which reduces the diffusion problem and allows the equilibrium of the diffusion rate to be reached quickly. This reduces the analysis time to about 1/10 compared to nitrogen adsorption, and the analysis time for CO_2 adsorption is several hours at most.

Krypton sorption at 77 K is suitable for very small sample volumes or samples with very small surface areas. The reason for this is that the saturation pressures of nitrogen and krypton at 77 K are very different in the subcooled state, 760 Torr for nitrogen and 2.6 Torr for krypton, which greatly improves the sensitivity of the probe system. As a result, at a given relative pressure and probe cell volume, the ratio of probe gas administered to probe gas adsorbed is much greater for nitrogen. That is, for gas adsorbed, nitrogen leaves more probe gas in the cell above the porous sample, by comparison, krypton. Therefore, the pressure change in the cell is small and difficult to measure for nitrogen, but large for krypton. Therefore, a sample with a small surface area, when probed with krypton compared to nitrogen, will have a larger change in pressure and be more reliable to measure.

The method described above can be applied to the evaluation of very small holes, but if you investigate large holes, you get no results and eventually fail. Macropores with diameters greater than 50 nm cannot be examined by volumetric gas adsorption methods. Instead, mercury intrusion porosimeter is used for pores with diameters from 0.04 to 400 microns. The smaller the cavity, the greater the pressure required to fill the mercury, and that pressure allows the size of the hole to be evaluated.

More exotic techniques include small-angle and wide-angle X-ray scattering (SAXS/WAXS) and positron annihilation lifetime spectroscopy (PALS). What these techniques have in common is that they can penetrate closed as well as open holes.

References

1. Langmuir, I. J. Am. *Chem. Soc.* 1936, 39, 1848–1906.
2. Brunauer, S., Deming, L.S., and Deming, W.E. *J. Am. Chem. Soc.* 1938, 60, 309–319.
3. Sing, R.A.W., Everet, K.S.W., and Haul, D.H. *Pure Appl. Chem.* 1985, 57, 603–619.
4. Thommes, M., Kaneko, K., Neimark, A.V., Olivier, J.P., Rodriguez-Reinoso, F., Rouquerol, J., and Sing, K.S.W. *Pure Appl. Chem.* 2015, 87, 1051–1069.
5. Cychosz, K.A. and Thommes, *M. Engineering*, 2018, 4, 559–566.
6. G. Mason, *J. Colloid Interface Sci.* 1982, 88, 36–46.
7. G. C. Wall and R. J. C. Brown, *J. Colloid Interface Sci.* 1981, 82, 141–149.
8. A. V. Neimark, *Dokl. Akad. Nauk SSSR* 1983, 273, 384–388.

Chapter 4

Gas Storage of CMPs

The design and synthesis of porous organic materials that have interesting physical and chemical characteristics have been discussed. Especially, conjugated microporous polymers (CMPs) have been designed and synthesized using simple and efficient procedures. Such materials have unique gas adsorption properties and can be used in gas separation and storage.

4.1 Natural Equilibrium

Climate change involves significant changes in long-term or short-term weather. The natural climate equilibrium is always stabilized by the national ecosystem and the overall cycle of carbon, water, and nitrogen. The carbon cycle consists of several geological, physical, chemical, and biochemical processes [1]. Emissions from industrial power plants that use coal and fossil fuels are one of the major sources of pollution and have a negative impact on plants, animals, and humans [3]. Therefore, to curb global warming, the development of sustainable energy sources is a priority [4–7]. However, many obstacles must be overcome to produce energy cleanly from green sources. So far, progress has been limited, and only a small amount of the energy needed can be produced from clean sources such as hydropower [8].

Conjugated Microporous Polymers
Atsushi Nagai

ISBN 978-981-5129-66-3 (Hardcover), 978-1-003-63652-6 (eBook)
www.jennystanford.com

4.2 Greenhouse Gas Emissions

Using clean energy sources is an effective way to solve environmental problems such as pollution and greenhouse gas (GHG) emissions from the use of fossil fuels. The use of porous organic materials as energy storage media is a potential solution because they can be used to capture and store clean fuels such as hydrogen (H_2) and methane (CH_4). In addition, porous organic materials can reduce GHG emissions [9, 10]. Because large amounts of greenhouse gases such as carbon dioxide (CO_2) and methane (CH_4) are currently emitted, solving the dangers and problems associated with global warming is a major issue. Such gases come from the exhaust gases of industrial plants and automobiles and are usually caught in the atmosphere. In addition, these gases cause floods and droughts, corrode natural gas pipes, and reduce the quality and capacity of natural gas [11, 12].

CMP was evaluated for gas storage and separation because of its unique structure and high porosity. Composed of lightweight elements, CMP has great potential for sound gas adsorption capacity and selectivity due to its low density. In addition, through rational unit design and structural control, the weight storage capacity can be improved. However, the use of expensive transition metals in many CMP syntheses may preclude their use in large-scale adsorption applications. Storage of H_2, CH_4, and CO_2 is the most widely studied area. H_2 and CH_4 are attracting attention because of their potential use as fuel. Especially, this chapter described hydrogen gas storage such as hydrogen storage and photocatalytic hydrogen evolution. CO_2 is a major greenhouse gas and causes global warming and ocean acidification, so adsorbents for this gas are attracting attention.

4.2.1 Carbon Dioxide Capture

Global warming is a major issue in the current scenario worldwide, which leads to climate change. Among the greenhouse gases, carbon dioxide plays a major role and its emissions rose to a historic high peak in recent years. Fossil fuel emissions result in about 92% of total CO_2 release because of expanding energy demand. Carbon dioxide is the enduring greenhouse gas, which absorbs hast. The other greenhouse gases like methane and nitrous oxide absorb more heat per molecule as compared to CO_2, which takes less. However, CO_2 is available in the atmosphere in large quantities and exits for a longer duration, as compared to other gases.

Through rational design as in Chapter 2, it is easy to obtain CMPs with high surface areas, suitable pore sizes, and CO_2-philic structures [13–15]. Oin and co-workers [16] synthesized a series of phenylene-based CMPs (A_6CMP-1~A_6CMP-7) by adjusting the molecular geometry of monomers. A_6CMP-6 with a suitable micropore size (0.46 nm) close to the kinetic radius of CO_2 exhibits the highest CO_2 capture capacity of 3.92 mmol g^{-1} at 273 K/1 bar.

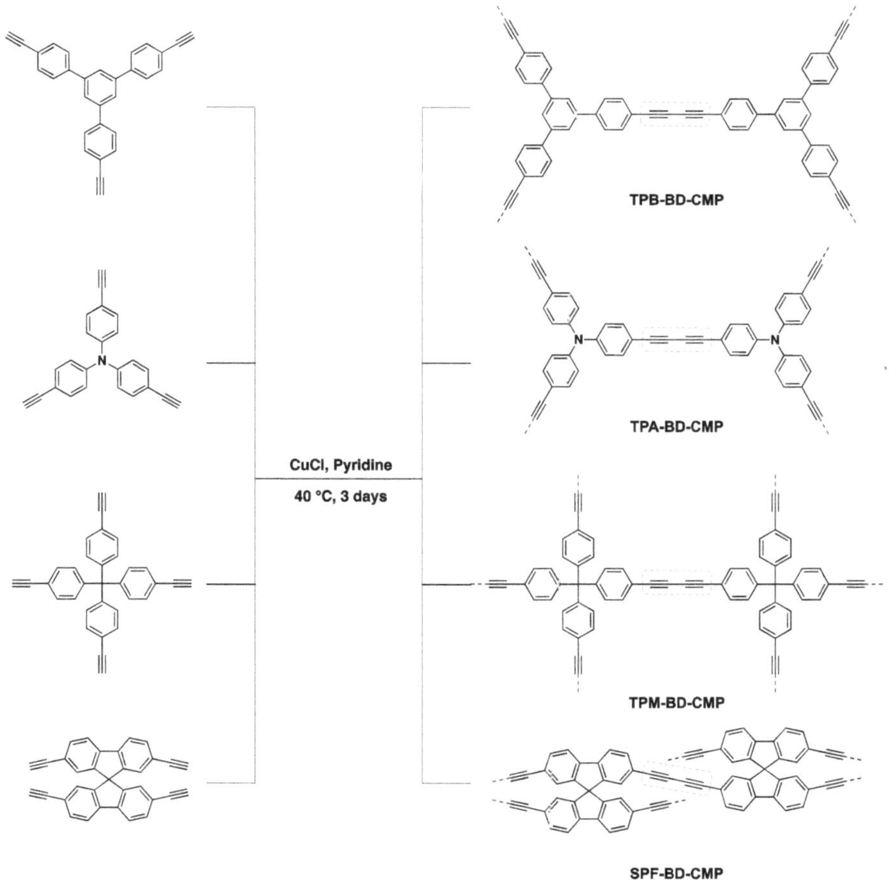

Figure 4.1 Synthetic routes for TPB-BD-CMP, TPA-BD-CMP, TPM-BD-CMP, and SPF-BD-CMP.

Heteroatoms can also be introduced into CMP for CO_2 capture. Xie et al., [17], prepared four butazinylene-based CMPs (TPB-BD-CMP, TPA-BD-CMP, TPM-BD-CMP, and SPF-BD-CMP) by the laser coupling reaction, of which the nitrogen-containing TPA-BD-CMP showed a higher CO_2 capture capacity of 3.28 mmol g^{-1} at 273 L and 1 bar, despite having a higher specific surface area than the non-nitrogenous TPB-BD-CMP with a similar structure (Figure 4.1). Ren et al. [18] prepared a series of porous carbons (PCM-1, -9NPCM-1) using CMP as a precursor. Compared with the non-nitrogen PCM-1, the nitrogen-doped NPCM-1 showed a much higher CO_2 capture capacity of 3.9 mmol g^{-1} at 273 K/l bar. Therefore, Xu et al. [19] reported three novel conjugated polymer networks were synthesized using a Sonogashira-Hagihara palladium-catalyzed cross-coupling reaction of 2,5-diborothiophene-3-carboxylic acid (DTCA) with aromatic ethynylbenzene to yield the heterocycle-based and carboxylic acid-functionalized CMP networks as SCMP-COOH@1, SCMP-COOH@2, and SCMP-COOH@3. The BET surface area of these porous polymers varied between 622 and 911 m^2 g^{-1}, and SCMP-COOH@1 showed an excellent CO_2 uptake capacity of 817 mg g^{-1} at 318 K under 60 bar pressure. Moreover, Faul et al. demonstrated that simple inorganic salts are used to tune N-containing CMPs synthesized by Buchwald-Hartwig (BH) cross-coupling reaction in the presence of salt catalysts [20]. The obtained poly(triphenylamines) (PTPAs) showed an improvement of BET surface area from 58 to 1152 m^2 g^{-1} where variations of the anion and cation sizes are used to fine-tune the surface area of PTPA, with the surface area showing a gradual decrease with an increase in the ionic radius of salts. This discovery shows that the Bristol-Xi'an Jiaotong (BXJ) approach provides a facile avenue to full control over CMPs. It opened a new low-cost and simple route for the control and optimization of the properties (surface area and porosity) and function of a broad class of N-rich CMPs, which is bound to impact energy storage materials and CO_2 capture (3.44 mmol g^{-1} at 273 K/l bar) and conversion (Figure 4.2). Thus, CMPs were regarded as promising CO_2 sorbents because of their low density, high specific surface areas, synthetic diversification, good physicochemical stability, tunable pore structure, and tailored functionality. There have been extensive investigations on CO_2 capture by CMPs. Table 4.1 lists some important related works.

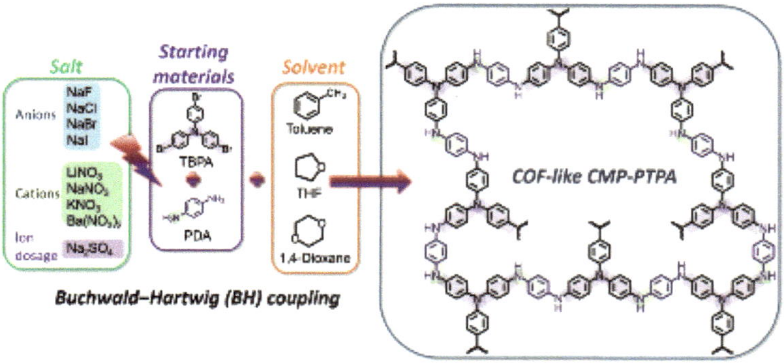

Figure 4.2 Synthetic route of the formation of salt-tunable PTPA networks. Reproduced from ref. [20] with permission from Willy.

Table 4.1 Surface area and CO_2 uptake capacity for CMPs

CMPs	**Brunauer-Emmett-Teller**	**CO_2 uptake capacity**			
	BET surface area ($m^2\,g^{-1}$)	**273 K (mmol g^{-1})**	**298 K (mmol g^{-1})**	**Pressure (bar)**	**Ref.**
A_6CMP-3	1042	3.17	3.17	1	[16]
A_6CMP-5	1110	3.43	3.43	1	[16]
A_6CMP-6	1115	3.61	3.61	1	[16]
TPA-BD-CMP	543	2.94	2.94	1	[17]
TPM-BD-CMP	1008	3.78	3.78	1	[17]
SPF-BD-CMP	833	3.25	3.25	1	[17]
PCM-1	630	3.6	-	1	[18]
NPCM-1	718	3.9	-	1	[18]
SCMP-COOH@1	911	96 (mg g^{-1}, 318 K)	-	60	[19]
SCMP-COOH@2	622	76 (mg g^{-1}, 318 K)	-	60	[19]
SCMP-COOH@3	820	84 (mg g^{-1}, 318 K)	-	60	[19]
PTPA	1114	3.44	-	1	[20]

The wide flexibility and structural tunability of the monomers allow CMP to be tuned for CO_2 capture. However, finding a low-cost way to prepare CMP on a large scale remains to be seen.

4.2.2 Hydrogen Gas Storage

The use of hydrogen requires safe and compact storage methods, but hydrogen is often stored in cylinders in energy-intensive ways, which carries additional risks. Therefore, it is best to convert hydrogen into a condensed state, and a direct method is to use a hydrogen adsorbent. As an energy storage medium, hydrogen has been attracting attention from research institutes and industry for the past 10 years. In order to produce hydrogen storage materials with good hydrogen storage properties under ideal hydrogen storage kinetics, the weak bond strength between hydrogen and adsorbent is an important obstacle. CMP materials are adsorption materials that have made significant progress in recent years [21–23]. The progress of the hydrogen storage capacity of CMPs in recent years is summarized in detail in Table 4.2. Most of the reported adsorption values were measured at a temperature of 77 K, which is well below the usual temperature range required for practical applications. Also, some reports give maximum adsorption values, while others report adsorption values at specific pressure (usually 1 bar). Therefore, it is difficult to synthesize all the reported values to make a judgment. Here, quintessential examples of strategies are discussed to improve hydrogen adsorption in recent years.

At a specific pressure and temperature (1.13 bar and 77.3 K), micropores rather than mesopores are the main reason for H_2 adsorption. High surface areas HCMP-1 and HCMP-2 were synthesized using homocoupling of 1,3,5-triethynylbenezne or 1,4-diethynylbenzene [24]. Although the apparent BEY surface area (827 $m^2\ g^{-1}$) of HCMP-2 is slightly smaller than that of HCMP-1 (842 $m^2\ g^{-1}$), the adsorption capacity of HCMP-2 (131 $cm^3\ g^{-1}$) on H_2 is higher than that of HCMP-1 (107 $cm^3\ g^{-1}$), which can be attributed to the higher micropore surface area and ultra-micropore volume observed in HCMP-2. Microwave-assisted rapid synthesis is a new synthesis technology, which has great industrial application potential in organic synthesis and material preparation and is conducive to enhancing the hydrogen adsorption capacity of organic microporous polymers. In 2010, microwave-enhanced ionothermal polymerization was used in the synthesis of porous polymer materials, which can in a few minutes obtain a high surface area (2390 m^2 g^{-1}) triazine-based CMPs (CTFs), and have a higher adsorption capacity of hydrogen (1.78 wt.%, 1 bar, and 77 K) [25]. It is worth noting that this method is a simple and efficient method for the preparation of highly efficient and economical polymer-based hydrogen adsorption materials.

In addition, lithium provides CMP with extremely high H_2 storage capacity, greatly advancing its application [26–28]. In 2016, under the guidance of computer simulation, methyllithium (MeLi) was designed as a new lithium dopant for preparing lithium-doped CMPs [28]. MeLi was designated as a cluster of H_2 molecules and naphthalene-containing CMP (N-CMP) as a matrix material to suppress the gathering effect of lithium. More notably, the strong interaction enables MeLi to be selectively anchored to the predesigned active sites, thereby achieving an optimal doping level. Thus, the hydrogen storage capacity of the CMP is greatly improved. MeLi-doped CMP (MeLi@N-CMP) improved the adsorption capacity of H_2 not only at low temperatures but also at near-normal temperatures. The hydrogen adsorption capacity of MeLi@N-CMP maintained an increase of 150% and 100% at 77 K, 80 bar (6.5 wt%), and 273 K, 80 bar (1.0 wt%), respectively.

Recently, Yang et al. designed and synthesized triazatruxene (TAT) with 3 coplanar indole moieties and further prepared high crosslinking poly (triazatruxene) (PTAT) using $FeCl_3$-catalyzed oxidative coupling polymerization [27]. The point-plane metal coordination interaction (cation-π) between the surface of the PTAT aromatic structure and the cation is beneficial to trap more lithium ions, and the trapped lithium ions are well dispersed on the PTAT surface to ensure sufficient Li^+ content in the PTAT. Lithium-doped CMP (Li^+-PTAT) can spontaneously adsorb hydrogen molecules by electrostatic charge quadrupole and charge-induced dipole interactions to obtain excellent hydrogen storage properties. Therefore, Li^+-PTAT showed higher hydrogen adsorption of 7.3 wt% (1 bar, 77 K) than PTAT of 1.9 wt% (1 bar, 77 K) under the same conditions. This adsorption value is higher than previously reported lithium-doped MOFs [29] and lithium-doped CMPs [27]. In addition, high hydrogen adsorption of 0.32 wt% at 1 bar and 273 K was maintained in Li^+-PTAT, indicating the possibility of application.

Furthermore, the development of selective hydrogen adsorption has attracted much attention. Membrane separation is a promising way to selectively adsorb hydrogen from mixed gases. Currently, the gas permeability of commercial films is low, and it is often necessary to prepare large-size films to maintain sufficient gas production [30], which obviously increases the cost of commercial applications. To improve the permeability and selectivity of the films, zeolites and MOFs with rigid skeletons and uniform pores were investigated extensively [31–34]. However, presently commercial films are mainly prepared from polymers with low permeability and high selectivity [35]. Therefore, a more ideal choice is to prepare MOPs with high permeability, good selectivity, and machinability. For example, the porous organic cages (POCs) prepared by Cooper's group are a class of MOP with good processability and show the potential for selective gas separation [35–37]. Polymers of intrinsic microporosity (PIMs) are a kind of MOPs with gas permeability, selectivity, and good solubility and machinability [30, 38–40]. CMPs have good chemical stability and can maintain functional stability under high pressure and temperature [41–43]. However, the extremely rigid framework of CMPs makes it difficult to reprocess. Therefore, the development of machinable, structurally stable CMPs with high gas selectivity and permeability is a major obstacle to be solved.

Recently, Loh et al. succeeded in synthesizing a solution-processable conjugated microporous thermoset (CMT) [44]. The CMT has excellent hydrogen permeability, solubility, and processability, and maintains a stable microporous structure and internal pore connectivity even at high temperatures and pressures. Because the internal pore connectivity of the CMT facilitates the redistribution of gas molecules, the CMT has a similar selective hydrogen adsorption capacity while achieving higher H_2 permeability than state-of-the-art ultra-pore PIMs [30, 40]. On the other hand, it is noteworthy that in addition to the solubility and processability of CMT, the centrally distributed pore size distribution is suitable for producing large-area film products.

Table 4.2 Summary of hydrogen storage in CMPs

CMPs	S_{set} (m^2 g^{-1})	V_{total} (cm^3 g^{-1})	P (bar)	T (K)	H_2 (wt.%)	Ref.
Li-CMP	795	1.61	1	77	6.1	[26]
POP-1	1031	0.641	60	77/87	2.78/2.31	[24]
POP-2	1013	0.712.	60	77/87	2.71/2.14	[24]
POP-3	1246	0.729	60	77/87	3.07/2.51	[24]
POP-4	1033	0.730	60	77/87	2.35/2.51	[24]
BLP-1(Cl)	1364	0.746	1	77	1.10	[45]
BLP-1(Br)	503	0.303	1	77	0.68	[45]
BLP-2(Cl)	1174	0.646	1	77	1.30	[45]
BLP-2(Br)	846	0.571	1	77	0.98	[45]
BLP-10(Cl)	924	0.67	1	77/87	1.3/0.9	[46]
PCTF-1	2235	1.56	1	77	1.86	[47]
PCTF-2	784	0.38	1	77	0.9	[47]
PhPOP-1	1050	0.69	1	77	2.23	[48]
PhPOP-2	160	0.27	1	77	1.03	[48]
PTAT	304	-	1	77	1.9	[27]
L^+-PTAT	281	-	1	77	7.3	[27]
CMP-G1	997	1.32	1/20	77	1.27/2.69	[49]
CMP-G2	786	0.87	1/20	77	1.01/2.14	[49]
HCMP-1	842	1.16	1.13	77.3	107 [a]	[24]
HCMP-2	827	1.35	1.13	77.3	131 [a]	[24]
p-PPF	269	0.20	45	77	67 [b]	[50]
m-PPF	229	0.15	45	77	56 [b]	[50]
SCMP-1	505	1	1	77	3.8 [c]	[25]
BILP-10	787	-	40	77	27.3 [d]	[51]

[a] H_2 uptake (cm^3 g^{-1})
[b] H_2 uptake (ml g^{-1})
[c] H_2 uptake (mmol g^{-1})
[d] H_2 uptake (g L^{-1})

4.2.3 CMPs for Photocatalytic Hydrogen Evolution

Photocatalytic technology has a number of important features [52–54], including: (i) it can operate under mild temperatures and pressures, and (ii) the parent body and intermediate products can be completely mineralized without secondary contamination. Hydrogen production by photocatalytic water splitting is regarded as a common technology to solve energy crises and environmental pollution [55–60]. Currently, many of the inorganic photocatalytic materials used are not conducive to further development because of complex preparation strategies, limited reserves, and low photocatalytic activity in visible light [61, 62]. From a sustainable development perspective, it is necessary to develop the synthesis of a variety of organic polymer photocatalysts that are metal-free, flexible in design, and tunable in structure and properties for photocatalytic hydrogen evolution [63–67]. Wang et al. present the basic theory of photocatalysis and the important factors affecting photocatalysis and discuss the development of new photocatalysis for inorganic materials, carbon-based materials, and semiconductor composites developed over the past 10 years [68]. It is also investigating and researching high-efficiency optical materials for converting solar energy into sustainable fuels. Finally, future developments in photocatalytic technology are discussed. CMP materials as photocatalysts have the following advantages [69–73]: (i) The expansion and opening of the p-conjugated structure favors the migration and separation of photocarriers.

(ii) High chemical and thermal stability provide excellent photocatalytic stability of CMP.

(iii) A wide range of raw materials can be used as synthetic monomers, including many economical and earth-rich raw materials.

(iv) A wide spectral absorption range is beneficial for improving the efficiency of light adsorption.

In 2016, Cooper's group reported the relative importance of junction shape, comonomer length, and degree of planarization to the photocatalytic hydrogen evolution rate of CMP and demonstrated the potential of CMP as a photocatalyst for hydrogen evolution [74], but the relevant design rules are not well understood. For example, it is necessary to investigate the catalytic efficiency of CMP materials with different surface areas, geometries, and band gaps, and to investigate the influence of substituent patterns and conjugate lengths on the photocatalytic performance relationship. Therefore, Jiang's research group synthesized three kinds of perylene-containing CMP (pr-CMP) by Suzuki reaction of hydroquinone as raw materials for photocatalytic hydrogen evolution [75]. As a result, it was found that the pore characteristics and band gap of Pr-CMP can be adjusted by changing the type of substituent and the connection position. Among these, the hydrogen evolution rate of the 1, 2, 4, and 5-bonded polymer (Pr-CMP-3) under UV irradiation was the highest (12.1 mol h^{-1}) among the three polymers, because Pr-CMP-3 has a high surface area, a wide light adsorption spectrum, and an extended π-conjugated structure. These results indicate that the bond shape is one of the important factors affecting the photocatalytic activity of CMP.

Subsequently, the effect of crosslinker length on the photocatalytic properties of hydrogen production has been reported. Jiang et al. reported a series of dibenzothiophene dioxide-containing CMP (DBTD-CMP) photocatalysts [76]. Among these, DBTD-CMP1, a photocatalyst with a short benzene bridge, showed the most active hydrogen evolution rate of 2460 mm h^{-1} g^{-1} without using a platinum catalyst in visible light. This may be due to the increased distortion of the polymer backbone and the reduced conjugation and planarity of the main molecular chain, which inhibits the transport and separation of photoinduced carriers and reduces photocatalytic activity. Furthermore, it is noteworthy that DBTD-CMP1 loaded with 3 wt.% Pt catalyst exhibited a significantly higher hydrogen evolution rate (9200 um h^{-1} g^{-1}).

A rigid pyrene unit is a common building block unit of CMP materials [77–79]. Jiang's group designed a series of donor-p-acceptor (D-p-A) CMP photocatalysts using pyrene, benzothiazole, and benzene (biphenyl) as the donor, acceptor, and p-crosslinker units, respectively, and compared the effects of the molecular structure and the ratio of donor to receptor on the photocatalytic properties of CMPs [77]. The donor unit, p-crosslinker unit, and acceptor unit are indispensable in enhancing the photocatalytic performance. Meanwhile, the ratio of donor to acceptor has a big impact on the photocatalytic performance of the CMPs, when the ratio of pyrene to benzothiadiazole is 9:2, the photocatalytic hydrogen evolution rate of the CMP (PyBT-2) is as high as 106 mmol h-1 under ultraviolet irradiation. Furthermore, they designed two CMP photocatalysts (PyDOBT-1 and PyDOBT-2) copolymerization of pyrene and dibenzothiophene-S,S-dioxide as building blocks, and investigated the effect of a binding mode of the results showed that the photocatalytic activity of PyDOBT-1 dibenzothiophene S,S-dioxide on photocatalytic hydrogen evolution [79]. The results showed that the photocatalytic activity of PyDOBT-2 with 3,7-connection mode was higher than that of PyDOBT-1 (Figure 4.3A) with 2,8-connection mode. The photocatalytic hydrogen production rate of "naked" PyDOBT-1 was 5697 mmol h−1g−1 under visible light irradiation, and the photocatalytic hydrogen production rate of Pt-load PyDOBT-1 reached a maximum of 12,986 mmol h−1g−1 (Figure 4.3B). The higher photocatalytic activity of PyDOBT-1 may be due to the 3,7-connection mode, which enhances the conjugated chain length and improves the coplanar property of the polymerized main chain to facilitate the transfer of the chain. Their work suggests that rational molecular design is the key to the development of effective CMP photocatalysts.

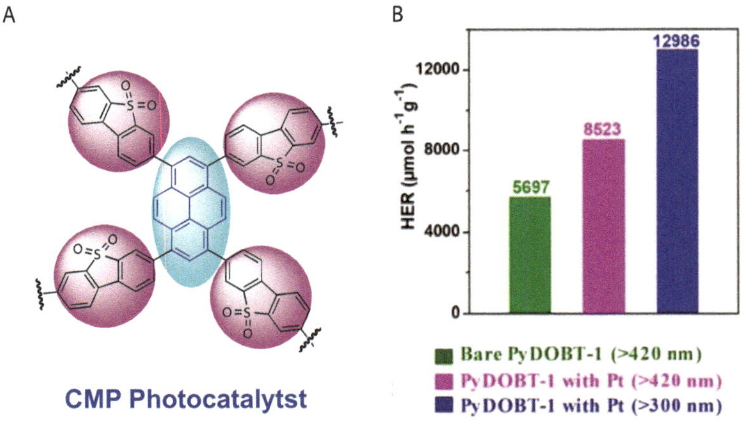

Figure 4.3 (A) Molecular structure of PyDOBT-1 and (B) hydrogen generation of the polymers under visible light [79].

Recently, Maji et al. synthesized a series of CMPs ($F_{0.1}$ CMP, $F_{0.5}$ CMP, $F_{2.0}$ CMP) with tetraphenylethylene (TPF) as donor and 9-fluorene (F) as acceptor, and tuned the optical properties of these CMPs by band gap engineering [80]. By increasing the F content, the emission color of the CMP material changed from green to red, the band gap of F_xCMP was adjusted from 2.8 eV to 2.1 eV, and the visible light absorption was improved. Among all CMPs, $F_{0.5}$ CMP (2.3 eV) showed the highest hydrogen evolution rate under broad-spectrum irradiation. Compared with the CMP material without F, the photocatalytic activity of D-A CMP with F was clearly improved. Interestingly, CMP showed emission at 540–580 nm, suggesting energy transfer at the TPF segment.

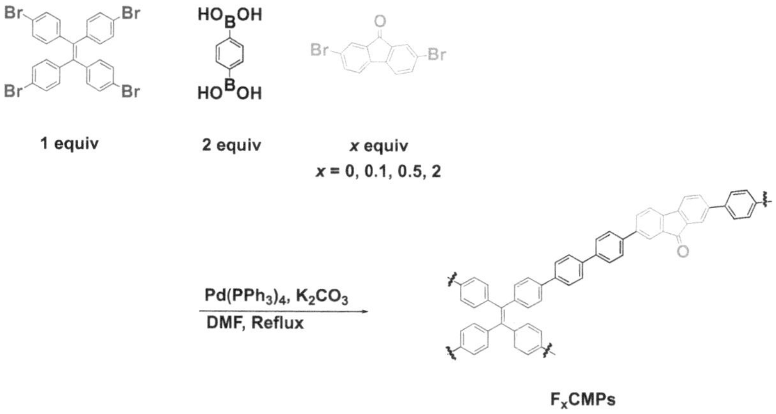

Figure 4.4 The preparation route of F_xCMPs.

As amorphous polymers, the effort of crystallization on the photocatalytic behavior of CMPs needs to be further discussed. In 2008, Cooper et al. targeted and investigated the difference in photocatalytic hydrogen production between benzobis(benzothiophene sulfone) moiety-based COF crystals and amorphous similar organic polymers [81]. Although the photocatalytic hydrogen evolution performance of COFs with ordered crystal structure is superior to that of their amorphous organic analogs in this case, this does not indicate that the photocatalytic activity of crystallized COFs is necessarily higher than that of amorphous CMPs. This is because crystallinity is not the only factor affecting photocatalytic activity [82–84]. Amorphous CMPs have much lower specific surface area and change transfer capacity, and some CMPs also have poor hydrophilic performance, which may result in reduced photocatalytic performance.

Besides the stability of photocatalysts should also be considered, CMP materials have been shown to have a more stable structure than the crystallized ordered COFs originally reported [85–88]. However, things are still changing, and researchers studying COF materials are exploring the use of more stable chemical bonds to build super-stable COF materials [89–91]. But, meanwhile, noteworthily, the amorphous properties of CMPs provide more design possibilities, and more possible multi-component CMP catalysts are constructed by using multi-step tandem reactions, which are difficult to be completed by COF materials.

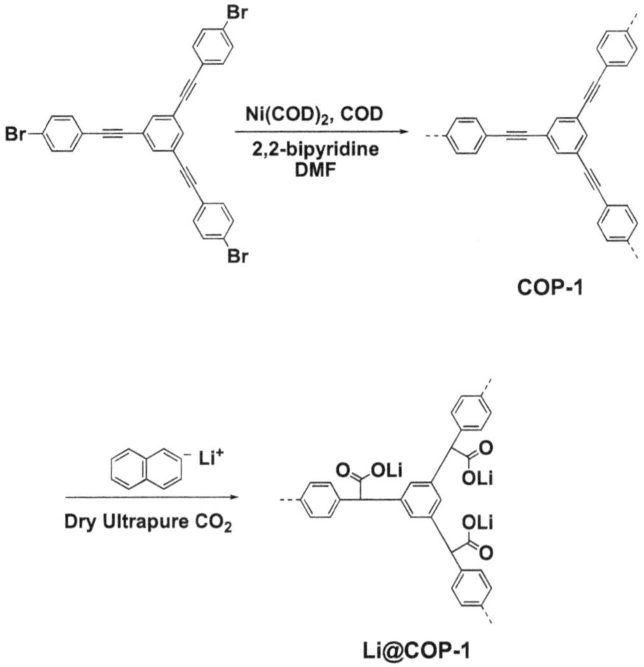

Figure 4.5 Schematic representation of the synthesis of Li@COP-1 [92].

4.2.4 Methane Gas Storage

Methane gas, the main component of natural gas, is an abundant energy storage resource. The target value for methane gas storage set by the Department of Energy (DOE) is 180 cm3 (STP) at 35 bar and 298 K. Organic porous materials are very promising because the heat of isothermal sorption of methane in carbonaceous materials is in the range appropriate for storage near ambient temperature. COP-1 (Figure 4.5) showed a CH4 adsorption capacity of 29 mg g–1 at 298 K and 18 bar. The addition of Li-ions increases the Li @ COP-1 to 39 mg g^{-1} [92]. Compared to metal-organic frameworks (MOF) and covalent or covalent organic frameworks (COF), the uptake capacity is still low. The highest reported volumes for COF (COF-102) [87] and MOF (MOF-210) [93] are 187 mg g^{-1} and 220 mg g^{-1}, respectively.

References

1. Abas N., Khan N. Carbon conundrum, climate change, CO_2 capture and consumptions. *J CO_2* 2014, 8, 39–48.
2. Zhang F., Xu L., Chen J., Chen X., Niu Z Lei, Li C., Zhao J. Chemical characteristics of PM2.5 during haze episodes in the urban of Fuzhou, China. *Particuology* 2013, 11(3), 264–272.
3. Diapouli E., Chaloulakou A., Spyrellis N. Levels of ultrafine particles in different microenvironments–implications to children exposure. *Sci Total Environ* 2007, 388(1–3), 128–136.
4. Xiao G., Wang X., Zhang J., Ni M., Gao X., Luo Z, Cen K. Granular bed filter: a promising technology for hot gas clean-up. *Powder Technol* 2013, 244, 93–99.
5. Heidenreich S. Hot gas filtration – a review. *Fuel* 2013, 104, 83–94.
6. Dunn S. *Hydrogen Futures: Toward a Sustainable Energy System.* International Energy Agency, Paris, 2001.
7. Riahi K., Roehr R.A., Schrattenholzer L., Miketa A. Technology clusters in sustainable development scenarios. *Progress Report of Environmental Issue Groups.* International Forum of the Collaboration Projects in Spring, Tokyo, 2001.
8. van den Berg AWC, Berg VD, Areán CO. Materials for hydrogen storage: current research trends and perspectives. *Chem Commun* 2008, 6, 668–681.
9. Férey G, Serre C, Devic T, Maurin G, Jobic H, Llewellyn PL, De Weireld G, Vimont A, Daturi M, Chang J-S. Why hybrid porous solids capture greenhouse gases? *Chem Soc Rev* 2011, 40(2), 550–562.
10. Wang S, Yan S, Ma X, Gong J. Recent advances in capture of carbon dioxide using alkali-metal-based oxides. *Energy Environ Sci* 4(10), 3805–3819.

11. Mastalerz M, Schneider MW, Oppel IM, Presly O. A salicylbisimine cage compound with high surface area and selective CO_2/CH_4 adsorption. *Angew Chem Int Ed* 2011, 50(5), 1046–1051.

12. D'Alessandro DM, Smit B, Long JR. Carbon dioxide capture: prospects for new materials. *Angew Chem Int Ed* 2010, 49, 6058–6082.

13. Xu Y, Chang D, Feng S, Zhang C, Jiang J. BODIPY-containing porous organic polymers for gas adsorption. *New J. Chem.* 2016, 40, 9415–9423.

14. Xu Y, Li Z, Zhang F, Zhuang X, Zeng Z, Wei J. New nitrogen-rich azo-bridged porphyrin-conjugated microporous networks for high performance of gas capture and storage. *RSC Adv.* 2016, 6, 30048–30055.

15. Wang S, Liu Y, Yu Y, Du J, Cui Y, Song X, Liang Z. Conjugated microporous polymers based on biphenylene for CO_2 adsorption and luminescence detection of nitroaromatic compounds. *New J. Chem.* 2018, 42, 9482–9487.

16. Qin L, Xu G, Yao C, Xu Y. Conjugated microporous polymer networks with adjustable microstructures for high CO_2 uptake capacity and selectivity. *Chem. Commun.* 2016, 52, 12602–12605.

17. Xie Z, Wei Y, Zhao X, Li Y, Ding S, Chen L. Facile construction of butadiynylene based conjugated porous polymers by cost-effective Glaser coupling. *Mater. Chem. Front.* 2017, 1, 867–872.

18. Xu Y, Wu S, Ren S, Ji J, Yue Y, Shen J. Nitrogen-doped porous carbon materials generated via conjugated microporous polymer precursors for CO_2 capture and energy storage. *RSC Adv.* 2017, 7, 32496–32501.

19. Qin, L., Xu, G.-J., Yao, C., Xu, Y.-H. Thiophene-based conjugated microporous polymers: preparation, porosity, exceptional carbon dioxide absorption and selectivity. *Polym. Chem.* 2016, 7, 4599–4603.

20. Chen, J., Yan, W., Townsend, E.J., Feng, J., Pan, L., Del Angel Hernandez, V., Faul, C.F.J. *Angew. Chem. Int. Ed.*, 2019, 38, 11715–11719.

21. Liao Y, Cheng Z, Trunk M, Thomas A. Targeted control over the porosities and functionalities of conjugated microporous polycarbazole networks for CO_2-selective capture and H_2 storage. *Polym Chem* 2017, 8, 7240–7247.

22. Chen Q, Luo M, Wang T, Wang J-X, Zhou D, Han Y, Zhang C-S, Yan C-G, Han B-H. Porous organic polymers based on propeller-like hexaphenylbenzene building units. *Macromolecules* 2011, 44, 5573–7.

23. Chen Q, Wang J-X, Yang F, Zhou D, Bian N, Zhang X-J, Yan C-G, Han B-H. Tetraphenylethylene-based fluorescent porous organic polymers: preparation, gas sorption properties and photoluminescence properties. *J Mater Chem* 2011, 21, 13554–60.

24. Jiang J-X, Su F, Niu H, Wood CD, Campbell NL, Khimyak YZ, Cooper AI. Conjugated microporous poly(phenylene butadiynylene)s. *Chem Commun* 2008, 486–488.

25. Zhang W, Li C, Yuan Y-P, Qiu Ll-G, Xie A-J, Shen Y-H, Zhu J-F. Highly energy- and time-efficient synthesis of porous triazine-based framework: microwave-enhanced ionothermal polymerization and hydrogen uptake. *J Mater Chem* 2010, 20, 6413–6415.

26. Li A, Lu R-F, Wang Y, Wang X, Han K-L, Deng W-Q. Lithium-doped conjugated microporous polymers for reversible hydrogen storage. *Angew Chem Int Ed* 2010, 49, 3330–3333.

27. Yang L, Ma Y, Xu Y, Chang G. Cation-π induced lithium-doped conjugated microporous polymer with remarkable hydrogen storage performance. *Chem Commun* 2019, 55, 11227–11230.

28. Xu D, Sun L, Li G, Shang J, Yang R-X, Deng W-Q. Methyllithium-doped naphthyl-containing conjugated microporous polymer with enhanced hydrogen storage performance. *Chemistry-A European Journal* 2016, 22, 7944–7949.

29. Mulfort KI., Hupp JT. Chemical reaction of metal-organic framework materials as a method to enhance gas uptake and binding. *J. Am. Chem. Soc.*, 2007, 129, 9604–9605.

30. Carte M., Malpass-Evans R., Croad M., Rogan Y., Jansen JC, Bernado P., Bazzerlli F., MckKeown, NB. An efficient polymer molecular sieve for membrane gas separation. *Science* 2013, 339, 303–307.

31. Rangnekar N, Mittal N, Elyassi B, Caro J, Tsapatsis M. Zeolite membranes – a review and comparison with MOFs. *Chem Soc Rev* 2015, 44, 7128–7154.

32. Jeon MY, Kim D, Kumar P, Lee PS, Rangnekar N, Bai P, Shete M, Elyassi B, Lee HS, Narasimharao K, Basahel SN, Al-Thabaiti S, Xu W, Cho HJ, Fetisov EO, Thyagarajan R, DeJaco RF, Fan W, Mkhoyan KA, Siepmann JI, Tsapatsis M. Ultra-selective high-flux membranes from directly synthesized zeolite nanosheets. *Nature* 2017, 543, 690–694.

33. Peng Y, Li Y, Ban Y, Jin H, Jiao W, Liu X, Yang W. Metal-organic framework nanosheets as building blocks for molecular sieving membranes. *Science* 2014, 346, 1356–1359.

34. Wang X, Chi C, Zhang K, Qian Y, Gupta KM, Kang Z, Jiang J, Zhao D. Reversed thermo-switchable molecular sieving membranes composed of two-dimensional metal-organic nano sheets for gas separation. *Nat. Commun* 2017, 8, 14460/1–10.

35. Song Q, Jiang S, Hasell T, Liu M, Sun S, Cheetham AK, Sivaniah E, Cooper AI. Porous organic cage thin films and molecular-sieving membranes. *Adv. Mater.* 2016, 28, 2629–2637.

36. Hasell T, Cooper AI. Porous organic cages: soluble, modular and molecular pores. *Nat Rev Mater* 2016, 1, 16053 /1–14.

37. Holst JR, Trewin A, Cooper AI. Porous organic molecules. *Nat. Chem.* 2010, 2, 915–920.

38. Song Q, Cao S, Zavala-Rivera P, Ping Lu L, Li W, Ji Y, Al-Muhtaseb SA, Cheetham AK, Sivaniah E. Photo-oxidative enhancement of polymeric molecular sieve membranes. *Nat Commun* 2013, 4, 1918 /1–9.

39. Jimenez-Solomon MF, Song Q, Jelfs KE, Munoz-Ibanez M, Livingston AG. Polymer nanofilms with enhanced microporosity by interfacial polymerization. *Nat Mater* 2016. 15. 760–767.

40. Rose I, Bezzu CG, Carta M, Comesaña-Gándara B, Lasseuguette E, Ferrari MC, Bernardo P, Clarizia G, Fuoco A, Jansen JC, Hart Kyle E, Liyana-Arachchi TP, Colina CM, McKeown NB. Polymer ultra permeability from the inefficient packing of 2D chains. *Nat. Mater.* 2017, 16, 932–937.

41. Dawson R, Cooper AI, Adams DJ. Nanoporous organic polymer networks. *Prog. Polym. Sci.* 2012, 37, 530–563.

42. Cooper AI. Conjugated microporous polymers. *Adv. Mater.* 2009, 21, 1291–1295.

43. Xu Y, Jin S, Xu H, Nagai A, Jiang D. Conjugated microporous polymers: design, synthesis and application. *Chem. Soc. Rev.* 2013, 42, 8012–8031.

44. Liu W, Jiang S-D, Yan Y, Wang W, Li J, Leng K, Japip S, Liu J, Xu H, Liu Y, Park I-H, Bao Y, Yu W, Guiver MD, Zhang S, Loh KP. A solution-processable and ultra-permeable conjugated microporous thermoset for selective hydrogen separation. *Nat. Commun.* 2020, 11, 1633 /1–7.

45. Reich TE, Jackson KT, Li S, Jena P, El-Kaderi HM. Synthesis and characterization of highly porous borazine-linked polymers and their performance in hydrogen storage application. *J Mater Chem* 2011, 21, 10629–10632.

46. Reich TE, Behera S, Jackson KT, Jena P, El-Kaderi HM. Highly selective CO_2/CH_4 gas uptake by a halogen-decorated borazine-linked polymer. *J Mater Chem* 2012, 22, 13524–13528.

47. Bhunia A, Vasylyeva V, Janiak C. From a supramolecular tetranitrile to a porous covalent triazine-based framework with high gas uptake capacities. *Chem Commun* 2013, 49, 3961–3963.

48. Sun C-J, Wang P-F, Wang H, Han B-H. All-thiophene-based conjugated porous organic polymers. *Polym Chem* 2016, 7, 5031–5038.

49. Zeng W, Zhang Y, Zhao X, Qin M, Li X, Jin W, Zhang D. One-pot synthesis of conjugated microporous polymers based on extended molecular graphenes for hydrogen storage. *Polymer (Guildf)* 2019, 174, 96–100.

50. Rao KV, Mohapatra S, Kulkarni C, Maji TK, George SJ. Extended phenylene microporous organic polymers with selective carbon dioxide adsorption. *J Mater Chem* 2011, 21, 12958–12963.

51. Rabbani MG, Sekizkardes AK, El-Kadri OM, Kaafarani BR, El-Kaderi HM. Pyrene-directed growth of nanoporous benzimidazole-linked nanofibers and their application to selective CO_2 capture and separation. *J Mater Chem* 2012, 22, 25409–25417.

52. Chong MN, Jin B, Chow CWK, Saint C. Recent developments in photocatalytic water treatment technology: a review. *Water Res* 2010, 44, 2997–3027.

53. Liu Y, Liu Z, Huang D, Cheng M, Zeng G, Lai C, Zhang C, Zhou C, Wang W, Jiang D, Wang H, Shao B. Metal or metal-containing nanoparticle MOF nanocomposites as a promising type of photocatalyst. *Coord Chem Rev* 2019, 388, 63–78.

54. Jiang L, Yuan X, Pan Y, Liang J, Zeng G, Wu Z, Wang H. Doping of graphitic carbon nitride for photocatalysis: a review. *Appl Catal B* 2017, 217, 388–406.

55. Liu J, Liu Y, Liu N, Han Y, Zhang X, Huang H, Lifshitz Y, Lee S-T, Zhong J, Kang Z. Metal-free efficient photocatalyst for stable visible water splitting via a two-electron pathway. *Science* 2015, 347, 970–974.

56. Zou W, Zhang L, Liu L, Wang X, Sun J, Wu S, Deng Y, Tang C, Gao F, Dong L. Engineering the Cu_2O–reduced graphene oxide interface to enhance photocatalytic degradation of organic pollutants under visible light. *Appl Catal B* 2016, 181, 495–503.

57. Yu H, Shi R, Zhao Y, Waterhouse GIN, Wu Ll-Z, Tung C-H, Zhang T. Smart utilization of carbon dots in semiconductor photocatalysis. *Adv Mater* 2016, 28, 9454–9477.

58. Zhang G, Lan Z-A, Wang X. Surface engineering of graphitic carbon nitride polymers with cocatalysts for photocatalytic overall water splitting. *Chem Sci* 2017, 8, 5261–5274.

59. Mahmood J, Li F, Jung S-M, Okyay MS, Ahmad I, Kim S-J, Park N, Jeong HY, Baek J-B. An efficient and pH-universal ruthenium-based catalyst for the hydrogen evolution reaction. *Nat Nanotechnol* 2017, 12, 441–446.

60. Chen X, Shen S, Guo L, Mao SS. Semiconductor-based photocatalytic hydrogen generation. *Chem Rev* 2010, 110, 6503–6570.

61. Li Y, Wang H, Xie L, Liang Y, Hong G, Dai H. MoS_2 nanoparticles grown on graphene: an advanced catalyst for the hydrogen evolution reaction. *J Am Chem Soc* 2011, 133, 7296–7299.

62. Sprick RS, Bonillo B, Clowes R, Guiglion P, Brownbill NJ, Slater BJ, Blanc F, Zwijnenburg MA, Adams DJ, Cooper AI. Visible-light-driven hydrogen evolution using planarized conjugated polymer photocatalysts. *Angew Chem Int Ed.* 2016, 55, 1792–1796.

63. Yu H, Shi R, Zhao Y, Bian T, Zhao Y, Zhou C, Waterhouse GIN, Wu Ll-Z, Tung C-H, Zhang T. Alkali-assisted synthesis of nitrogen deficient graphitic carbon nitride with tunable band structures for efficient visible-light-driven hydrogen evolution. *Adv Mater* 2017, 29, 1605148 /1–7.

64. Cheng W, Su H, Tang F, Che W, Huang Y, Zheng X, Yao T, Liu J, Hu F, Jiang Y, Liu Q, Wei S. Synergetic enhancement of plasmonic hot-electron injection in Au cluster-nanoparticle/C_3N_4 for photocatalytic hydrogen evolution. *J Mater Chem A* 2017, 5, 19649–19655.

65. Wang Y, Silveri F, Bayazit MK, Ruan Q, Li Y, Xie J, Catlow CRA, Tang J. Bandgap engineering of organic semiconductors for highly efficient photocatalytic water splitting. *Adv Energy Mater* 2018, 8, 1801084 /1–10.

66. Yu Y, Yan W, Wang X, Li P, Gao W, Zou H, Wu S, Ding K. Surface engineering for extremely enhanced charge separation and photocatalytic hydrogen evolution on g-C_3N_4. *Adv Mater* 2018, 30, 1705060 /1–8.

67. Kong D, Zheng Y, Kobielusz M, Wang Y, Bai Z, Macyk W, Wang X, Tang J. Recent advances in visible light-driven water oxidation and reduction in suspension systems. *Mater Today* 2018, 21, 897–924.

68. Vyas VS, Lau VW-h, Lotsch BV. Soft photocatalysis: organic polymers for solar fuel production. *Chem Mater* 2016, 28, 5191–5204.

69. Li L, Cai Z, Wu Q, Lo W-Y, Zhang N, Chen LX, Yu L. Rational design of porous conjugated polymers and roles of residual palladium for photocatalytic hydrogen production. *J Am Chem Soc* 2016, 138, 7681–7686.

70. Han XL, Xu MY, Yang S, Qian J, Hua DB. Acetylcysteine-functionalized microporous conjugated polymers for potential separation of uranium from radioactive effluents. *J Mater Chem A* 2017, 5, 5123–5128.

71. Wang F, Ren F, Mu P, Zhu ZQ, Sun HX, Ma CH, Xiao CH, Liang WD, Chen LH, Li A. Hierarchical porous spherical-shaped conjugated microporous polymers for the efficient removal of antibiotics from water. *J Mater Chem A.* 2017, 5, 11348–11356.

72. Liras M, Iglesias M, Sanchez F. Conjugated microporous polymers incorporating bodily moieties as light-emitting materials and recyclable visible-light photocatalysts. *Macromolecules* 2016, 49, 1666–1673.

73. Sprick RS, Bonillo B, Sachs M, Clowes R, Durrant JR, Adams DJ, Cooper AI. Extended conjugated microporous polymers for photocatalytic hydrogen evolution from water. *Chem Commun* 2016, 52, 10 0 08–11.

74. Xu Y, Mao N, Feng S, Zhang C, Wang F, Chen Y, Zeng J, Jiang J-X. Perylene–containing conjugated microporous polymers for photocatalytic hydrogen evolution. *Macromol Chem Phys* 2017, 218, 170 0 049 /1–9.

75. Wang Z, Yang X, Yang T, Zhao Y, Wang F, Chen Y, Zeng JH, Yan C, Huang F, Jiang J-X. Dibenzothiophene dioxide based conjugated microporous polymers for visible-light-driven hydrogen production. *ACS Catal* 2018, 8, 8590–8596.

76. Wang Z, Yang X, Yang T, Zhao Y, Wang F, Chen Y, Zeng JH, Yan C, Huang F, Jiang J-X. Dibenzothiophene dioxide based conjugated microporous polymers for visible-light-driven hydrogen production. *ACS Catal* 2018, 8, 8590–8596.

77. Xu Y, Mao N, Zhang C, Wang X, Zeng J, Chen Y, Wang F, Jiang J-X. Rational design of donor-pi-acceptor conjugated microporous polymers for photocatalytic hydrogen production. *Appl Catal B* 2018, 228, 1–9.

78. Xu YF, Zhang C, Mu P, Mao N, Wang X, He Q, Wang F, Jiang JX. Tetra-armed conjugated microporous polymers for gas adsorption and photocatalytic hydrogen evolution. *Sci China Chem* 2017, 60, 1075–1083.

79. Zhao Y, Ma W, Xu Y, Zhang C, Wang Q, Yang T, Gao X, Wang F, Yan C, Jiang J-X. Effect of linking pattern of dibenzothiophene-s,s-dioxide-containing conjugated microporous polymers on the photocatalytic performance. *Macromolecules* 2018, 51, 9502–9508.

80. Mothika VS, Sutar P, Verma P, Das S, Pati SK, Maji TK. Regulating charge transfer in conjugated microporous polymers for photocatalytic

hydrogen evolution. *Chemistry-A European Journal* 2019, 25, 3867–3874.

81. Wang X, Chen L, Chong SY, Little MA, Wu Y, Zhu W-H, Clowes R, Yan Y, Zwijnenburg MA, Sprick RS, Cooper AI. Sulfone-containing covalent organic frameworks for photocatalytic hydrogen evolution from water. *Nat Chem* 2018, 10, 1180–1189.
82. Wang H, Wang H, Wang Z, Tang L, Zeng G, Xu P, Chen M, Xiong T, Zhou C, Li X, Huang D, Zhu Y, Wang Z, Tang J. Covalent organic framework photocatalysts: structures and applications. *Chem Soc Rev* 2020, 49, 4135–4165.
83. Slater AG, Cooper AI. Function-led design of new porous materials. *Science* 2015, 348, 8075/1–10.
84. Wan S, Gándara F, Asano A, Furukawa H, Saeki A, Dey SK, Liao L, Ambrogio MW, Botros YY, Duan X, Seki S, Stoddart JF, Yaghi OM. Covalent organic frameworks with high charge carrier mobility. *Chem Mater* 2011, 23, 4094–4097.
85. Côté AP, Benin AI, Ockwig NW, Keeffe M, Matzger AJ, Yaghi OM. Porous, crystalline, covalent organic frameworks. *Science* 2005, 310, 1166–1170.
86. Jiang J-X, Su F, Trewin A, Wood CD, Campbell NL, Niu H, Dickinson C, Ganin AY, Rosseinsky MJ, Khimyak YZ, Cooper AI. Conjugated microporous poly(aryleneethynylene) networks. *Angew Chem Int Ed* 2007, 46, 8574–8578.
87. El-Kaderi HM, Hunt JR, Mendoza-Cortes JL, Cote AP, Taylor RE, O'Keeffe M, Yaghi OM. Designed synthesis of 3D covalent organic frameworks. *Science* 2007, 316, 268–272.
88. Weber J, Thomas A. Toward stable interfaces in conjugated polymers: microporous poly(p-phenylene) and poly(phenyleneethynylene) based on a spirobifluorene building block. *J Am Chem Soc* 2008, 130, 6334–6335.
89. Chandra S, Kandambeth S, Biswal BP, Lukose B, Kunjir SM, Chaudhary M, Babarao R, Heine T, Banerjee R. Chemically stable multilayered covalent organic nanosheets from covalent organic frameworks via mechanical delamination. *J Am Chem Soc* 2013, 135, 17853–17861.
90. Chandra S, Kandambeth S, Biswal BP, Lukose B, Kunjir SM Biswal BP, Kandambeth S, Chandra S, Shinde DB, Bera S, Karak S, Garai B, Kharul UK, Banerjee R. Pore surface engineering in porous, chemically stable covalent organic frameworks for water adsorption. *J Mater Chem A* 2015, 3, 23664–23669.

91. Guan X, Li H, Ma Y, Xue M, Fang Q, Yan Y, Valtchev V, Qiu S. Chemically stable polyarylether-based covalent organic frameworks. *Nat Chem* 2019, 11, 587–594.

92. Z. H. Xiang, D. P. Cao, W. C. Wang, W. T. Yang, B. Y. Han and J. M. Lu, *J. Phys. Chem. C*, 2012, 116, 5974–5980.

93. H. Furukawa, N. Ko, Y. B. Go, N. Aratani, S. B. Choi, E. Choi, A. O. Yazaydin, R. Q. Snurr, M. O'Keeffe, J. Kim, O. M. Yaghi, *Science*, 2010, 329, 424–428.

Chapter 5

Heterogeneous CMP's Catalyst Applications

Conjugated micropore polymer (CMP) is a type of cross-linked polymer that combines permanent micropores with a π-conjugated framework and has a three-dimensional (3D) network. By a bottom-up strategy, catalyst sites can be introduced directly into the framework to produce heterogeneous CMP catalysts. The heterogeneous CMP catalyst can obtain higher activity, stability, and selectivity than the homogeneous catalyst. In addition, the CMP catalyst is easy to isolate and recycle. This chapter focuses on CMP as an interesting platform for developing various heterogeneous catalysts that are highly efficient and recyclable in organic reactions. The design, synthesis, and structure of these CMP catalysts are also discussed.

5.1 Introduction

As discussed in the characteristics of the previous chapter, the diversity of building blocks is one of the most distinctive features of CMPs. Building blocks applied to CMP materials range from simple phenyl units to the like. By appropriately selecting building blocks and crosslinking strategies, CMP can fine-tune its porous structure and optimize its properties and skeleton. As a result, CMP has

Conjugated Microporous Polymers
Atsushi Nagai

ISBN 978-981-5129-66-3 (Hardcover), 978-1-003-63652-6 (eBook)
www.jennystanford.com

been applied in a wide range of fields. This chapter introduces the significant progress made by CMP in heterogeneous catalysis over the past few years. In a bottom-up strategy, heterogeneous CMP catalysts can be produced by introducing catalytic sites directly into the backbone. By homogeneously distributing catalyst sites with high surface area and loading, these heterogeneous CMP catalysts have potentially high activity. Since the CMP catalyst is insoluble, it is easy to isolate and recycle. Thus, CMP catalysts have attracted attention as an interesting platform for developing various heterogeneous catalysts that are highly efficient and recyclable in organic reactions. The design, synthesis, and structure of these CMP catalysts are also discussed in this chapter.

5.2 Conjugated Microporous Polymers for Heterogeneous Catalysis

5.2.1 Heterogeneous Transition-Metal Catalysis

Transition-metal catalysts play an important role in organic reactions. Organic coupling reactions involving transition metal catalysts, such as the Suzuki, Heck, and Negishi reactions, have been recognized because of the 2010 Nobel Prize. However, many homogeneous transition-metal catalysts were expensive and difficult to recycle from products, which limited their industrial application. Therefore, it is important to select an appropriate catalyst support. CMP with adjusted porosity and functionality can be a reasonable alternative because of its high physicochemical stability.

5.2.2 Heterogeneous Ir-Catalyzed Reductive Amination

Cooper et al. prepared a type of metal-organic conjugated microporous polymer (CMP-1) based on cyclometalated Ir complexes by a simple and versatile strategy, including direct metal introduction by copolymerization and metalation after synthesis [1]. In comparison with the metal node of the MOF [2] as a heterologous catalyst, CMP-1 showed high catalytic activity for reductive amination reactions on a wide range of substrates (Figure 5.1). The reductive amination activity of the heterogeneous catalyst CMP-1 is comparable to that of the homogeneous catalyst.

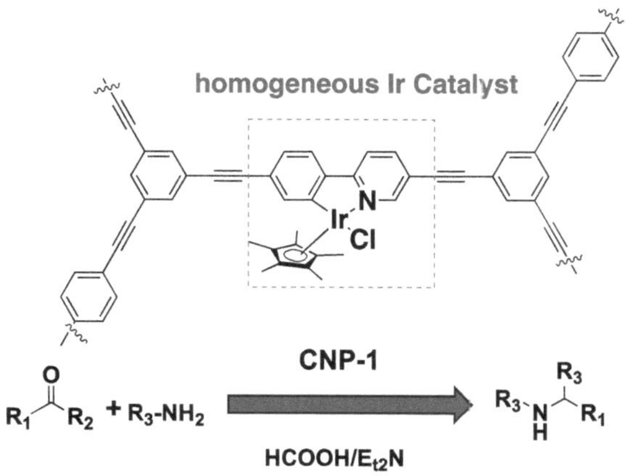

Figure 5.1 Reactive amination catalyzing by CMP-1.

5.2.3 CMPs for Heterogeneous Fe-Catalyzed Oxidation of a Sulfide and Epoxidation of Olefins

Because porphyrin and related metalloporphyrin derivatives facilitate many important biological processes, porphyrins in catalysis have been extensively and deeply studied over the past decades [3]. Previously, heterogeneous porphyrin catalysts have been developed by immobilization in porous solids, such as metal-organic frameworks (MOFs) [4] and polymeric materials [5].

Surprisingly, Jiang's group developed an iron (III) porphyrin-based CMP (Figure 5.2, CMP-2) with a large surface area using an iron porphyrin derivative as a building block and a catalytic species by a Suzuki coupling reaction [6]. The CMP-2 catalyst functions as a heterogeneous catalyst to activate oxygen molecules and selectively converts sulfides to the corresponding sulfoxides at room temperature. On the other hand, metal-organic frameworks (MOF) based on Sc metal sources have also been developed for this reaction, but stoichiometric amounts of materials are required [7]. When thioanisole (1 a) was used as a substrate, the reaction rate (TON) and selectivity increased to 97320 and 99/1, respectively (Scheme 5.1, Formula 1). Subsequently, the CMP-2 catalyst was also found to exhibit outstanding catalytic performance for selective epoxidation of trans-stilbene at room temperature (Scheme 5.1, Formula 2) [8]. The catalytic activity was demonstrated to increase in direct proportion to the surface area of CMP-2. When monomeric Fe-porphyrin was used as a catalyst instead of CMP-2, both conversion and selectivity were clearly reduced, indicating that porous CMP-2 was superior to monomeric Fe-porphyrin catalyst. As a control for CMP-2, linear CMP-3 was synthesized and showed poor catalytic performance like that of monomeric Fe-porphyrin complexes.

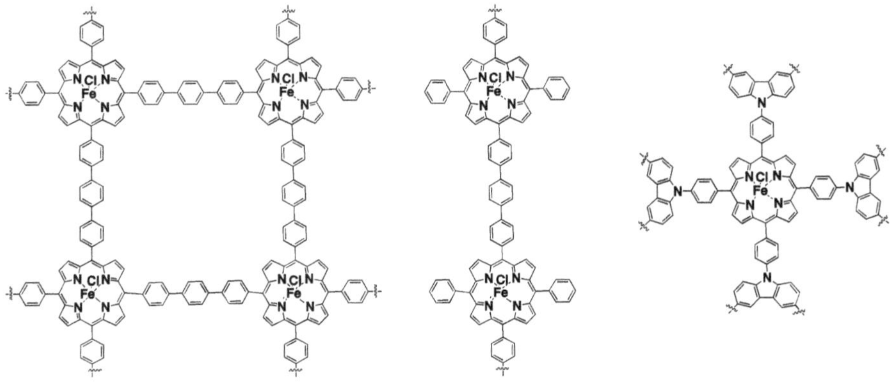

Figure 5.2 Structures of designed Fe-porphyrins (CMP-2, CMP3, and CMP-4).

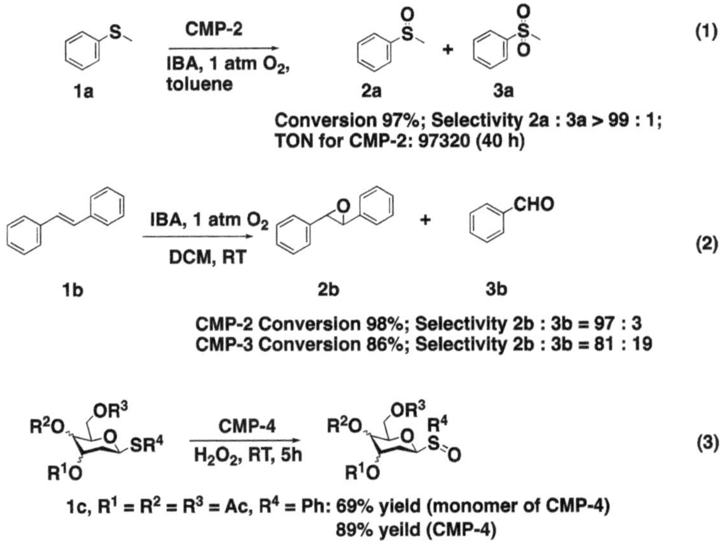

Scheme 5.1 Catalytic oxidation of sulfides and stilbenes over Fe-porphyrin-based CMP.

In another example, an iron porphyrin site was constructed in a conjugated porous framework (CMP-4) by a carbazole-based oxidative coupling reaction [9]. The resulting CMP-4 proved to be an effective heterogeneous catalyst for the selective oxidation of the corresponding thioglycosides to form glycosyl sulfoxides and showed robust recycling capability. Furthermore, the catalytic activity of CMP-4 (89% yield) was superior to that of the corresponding monomer (Yield 69%, Scheme 5.1, Formula 3).

5.2.4 Heterogeneous Ru-Catalyzed Asymmetric Hydrogenation

Li's group was inspired to introduce chiral organometallic catalysts into the structure of CMP. Such chiral CMPs not only function as heterologous catalysts with fine-tunable structures, but also prevent the formation of dimers and trimers, which are sometimes regarded as deactivated species because of their natural topological structure (Figure 5.3 a). Based on this concept, they reported a series of chiral CMPs with tunable pore size distribution and BET-specific surface area (Figure 5.3 b) [10]. In this design, chiral (R)-BINAP ligands

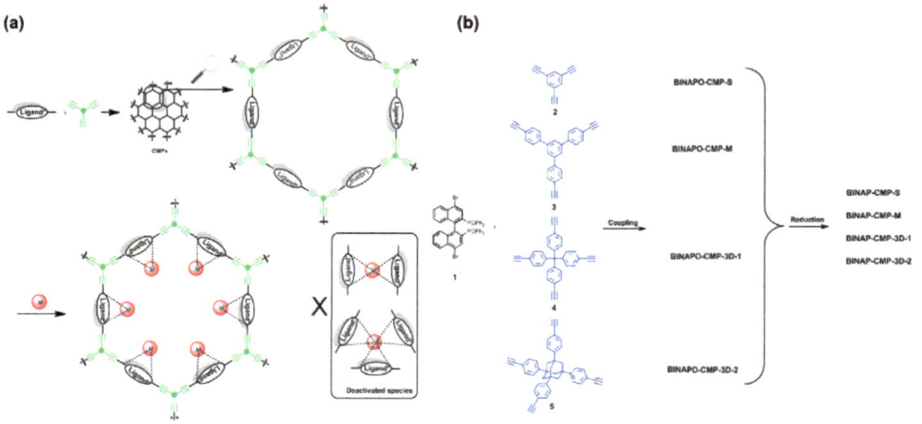

Figure 5.3 (a) The design of chiral CMPs and (b) synthetic route for the BINAP-CMPs [10].

were incorporated into the three-dimensional framework of CMP by the Sonogashira-Hagiwara reaction using alkynes of different lengths. Compared with other BINAP-based CMPs, CMP-5 achieved high enantioselectivity in Ru-catalyzed asymmetric hydrogenation of β-keto esters. This may be due to the large BET surface area and micropore volume (Table 5.1). For Ru/CMP-5d, no significant decrease in activity was observed after three cycles, but the *ee* value decreased from 94 to 90% (reaction time: 24 h). Furthermore, the CMP-5d-supported Ir species showed higher activity in the asymmetric hydrogenation of quinaldine than the homogeneous Ir/BNAP system.

Table 5.1 Reactivity comparison experiment and specific area (S_{BET}) of CMPs based on BINAP

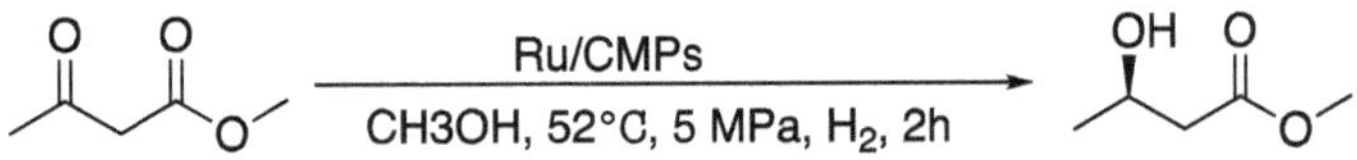

Entry	Catalyst	S_{BET} ($m^2\ g^{-1}$)	Conversion (%)	*ee* (%)
1	Ru/CMP-5a	391	26	67
2	Ru/CMP-5b	407	27	71
3	Ru/CMP-5c	474	24	72
4	Ru/CMP-5d	509	30	91

5.2.5 CMPs for Heterogeneous Ag-Catalyzed Heterocyclization

In 2015, Viela's group synthesized sulfur-containing CMP-7 by modifying CMP-6 using thiol-ene chemistry (Scheme 5.2) [11]. Ag/CMP-7 was prepared by impregnating $AgNO_3$ into an aqueous

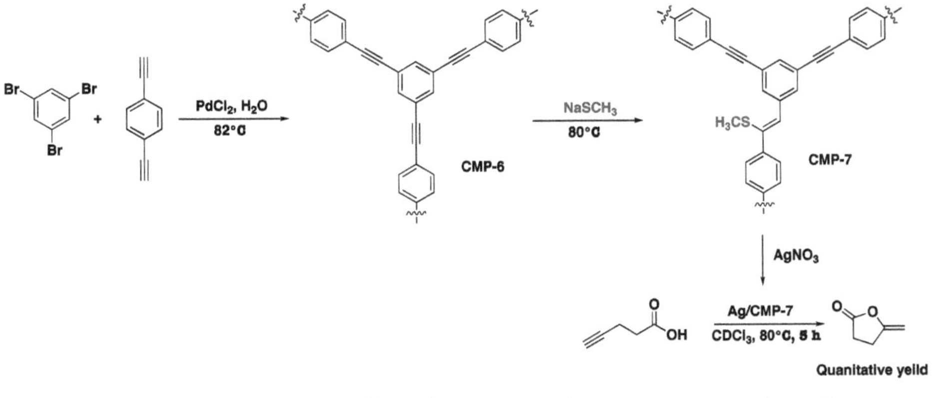

Scheme 5.2 Synthesis of Ag/CMP-7 as heterogenous catalyst for heterocyclization of 4-pentyroic acid.

solution of CMP-7 without light. It was used in the cyclization of 4-pemetymnoic acid and converted to 5-methylenedihydrofuran-2(3*H*), one with excellent yield. For comparison, the $AgNO_3$ salt as a homogeneous catalyst gave a very low conversion in the cyclization of 4-pentenoic acid. Furthermore, Ag/CMP-7 functioned without loss of efficiency, but the Ag content of the heterogeneous catalyst remained unchanged after 5 runs.

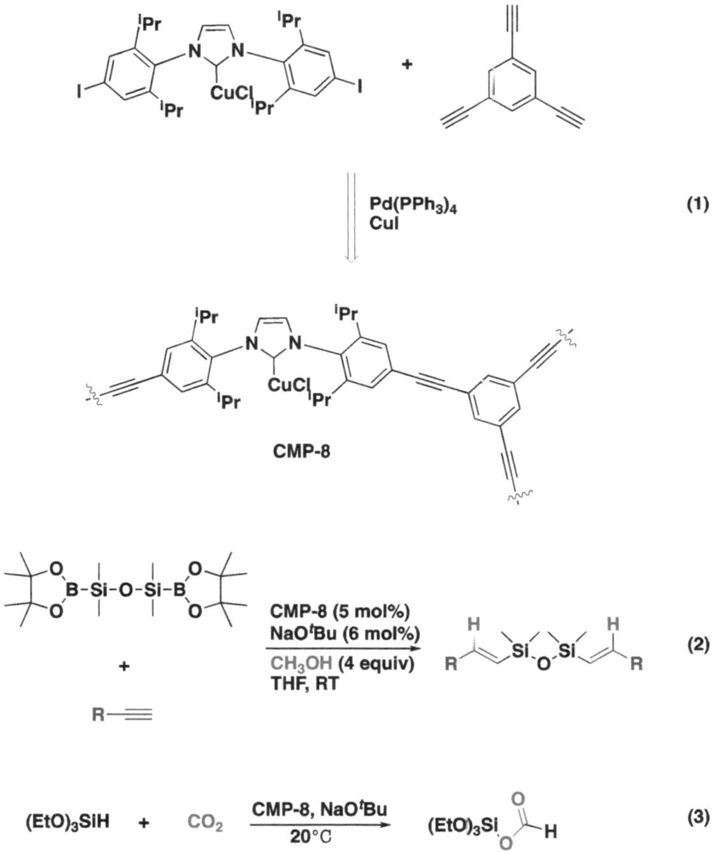

Scheme 5.3 Syntheisis of CMP-8 and its application in hydrosilylation of terminal alkyenes and COs.

5.2.6 Heterogeneous Cu-Catalyzed Hydrosilylation of Terminal Alkynes and CO_2

Zhou et al. developed a homogeneous catalyst IPr (CuCl) effective for the hydrosilylation of terminal alkynes with boryldisiloxanes

[12]. They succeeded in synthesizing heterogeneous (NHC) CuCl (CMP-8) by immobilizing homogeneous IPr (CuCl) on CMP by the Sonogashira coupling reaction (Scheme 5.3, Formula 1) of 1,3,5-triethynylbenezne with iodo-NHC-CuCl [13]. CMP-8 showed high activity and selectivity for hydrosilylation of a wide range of terminal alkynes, comparable to homogeneous IPr (CuCl) (Scheme 5.3, Formula 2). In addition to its high activity and selectivity, CMP-8 showed good stability and proved to be an effective catalyst for the hydrosilylation reaction of CO_2 under mild conditions (Scheme 5.3, Formula 3).

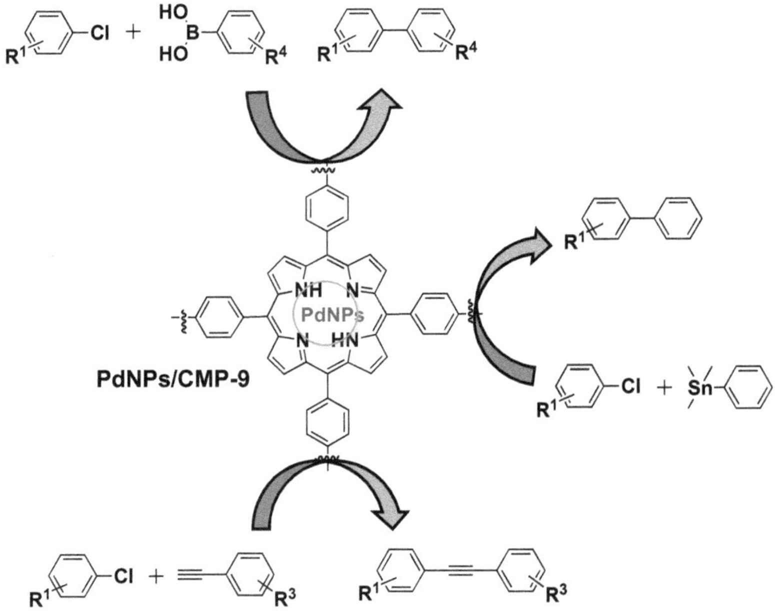

Figure 5.4 PdNPs/CMP-9 as heterogeneous catalyst for carbon-carbon bond coupling reactions.

5.2.7 Heterogeneous Pd-Catalyzed Carbon-Carbon Bonding Coupling Reactions

Jiang et al. designed a new type of organically conjugated microporous polymer CMP-9 by Yamamoto coupling polycondensation of 5, 10, 15, and 20-tetrakis(4'-bromophenyl) porphyrin [14]. The PdNPs/

CMP-9 was obtained by impregnating the CMP-9 dispersed in THF with $Pd(NO_3)_2$ and then reducing it under H_2 flow. The resulting heterologous catalyst, PdNPs/CMP-9, exhibited excellent catalytic activity in a series of carbon-carbon bonding reactions, including the Suzuki, Sonogashira, and Stille coupling reactions using unreacted aryl chlorides as substrates (Figure 5.4). Furthermore, the crystal structure and catalytic activity of PdNPs in CMP-9 were retained after 10 cycles by XRD measurements and repeated experiments, respectively.

Recently, Zhan's group developed a series of novel pyridine-type ligands containing the C ≡ C (Figure 5.5, L1–L5) [15]. Pyridine-type ligands (L1–L4) showed good l/b selectivity in the selective oxidation Heck reaction of 1-octene, but a moderate regioisomer ratio (Table 5.2., Entry 5).

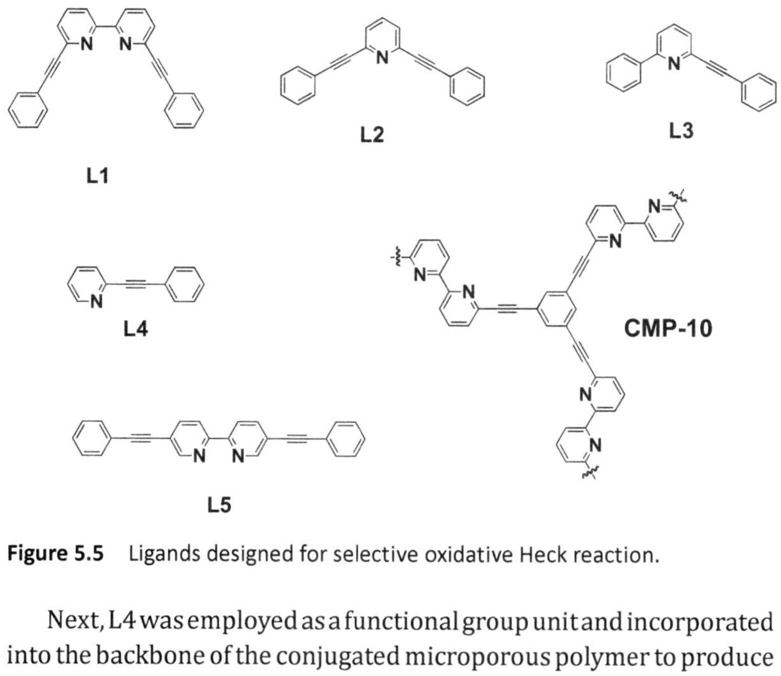

Figure 5.5 Ligands designed for selective oxidative Heck reaction.

Next, L4 was employed as a functional group unit and incorporated into the backbone of the conjugated microporous polymer to produce CMP-10 (Figure 5.5). Both SEM and TEM images show that CMP-10 has a hierarchical porosity (Figure 5.6). The obtained heterogeneous catalyst composed of CMP-10 and Pd (OAc)2 (Pd/CMP-10) exhibited excellent selectivity for the oxidative heck reaction of olefins without heteroatoms such as 1-octene (Table 5.2., Entry 6), indicating that CMP-10 outperformed monomeric pyridine-type ligands in controlling regioselectivity. The high selectivity for the oxidative Heck reaction can be attributed to the porous structure in the polymer and the bifunctional ligand features (consisting of C ≡ C and pyridine-*N* sites with appropriate coordination geometry angles). Furthermore, DFT calculations showed that the formation of linear isomers was much more favorable than that of branched isomers, both kinetically and thermodynamically, when homogeneous L4 was used, which was in good agreement with the experimental results.

Interestingly, the thiophene-type ligand containing the C ≡ C group was also found to promote the selective oxidative Heck reaction [16]. CMP-11 as a heterogenous ligand offered the corresponding product in 73% yield and an excellent ratio of ln/br (ln/br = 30:1), which was much better than white's catalyst (ln/br = 8:1, r. r. < 1:1) [17].

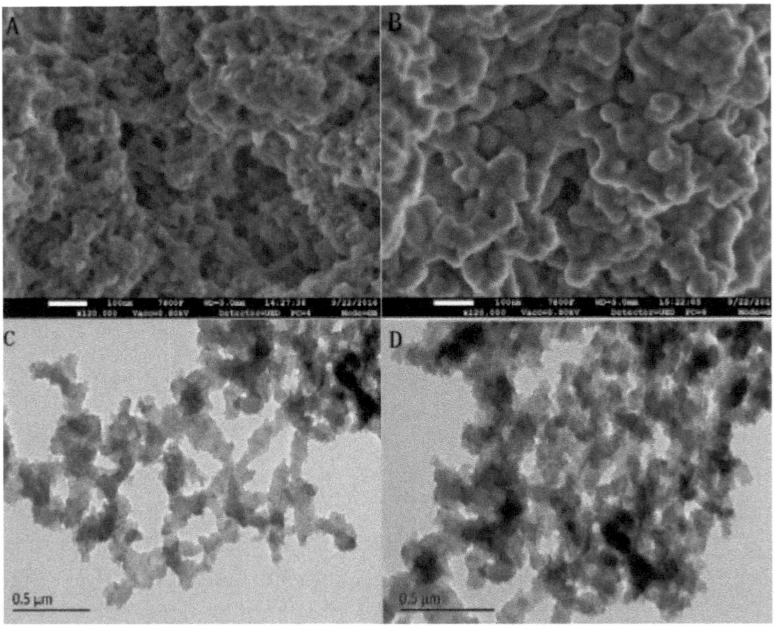

Figure 5.6 SEM of CMP-10 (A) and Pd/CMP-10 (B); TEM of CMP-10 (C) and Pd/CMP-10 (D) [15].

Table 5.2 Impact of different ligands on selective oxidative Heck[a]

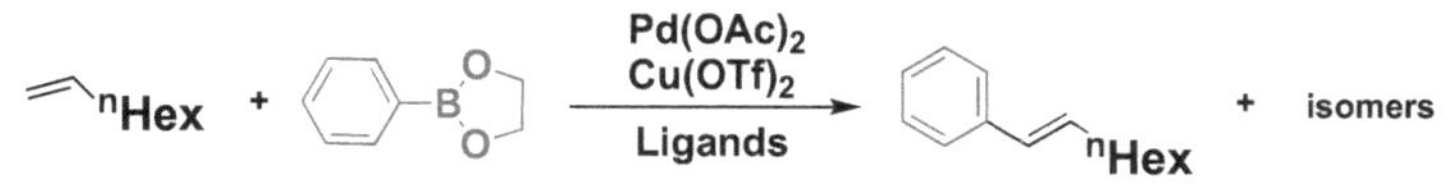

Entry	Ligands	Yield (%)	l:b[b]	r:r[c]
1	**L1**	41	12:1	2.1:1
2	**L2**	65	15:1	3.9:1
3	**L3**	84	20:1	4.4:1
4	**L4**	65	21:1	4.3:1
5	**L5**	88	2.5:1	1.3:1
6	**CMP-10**	76	70:1	30:1

a: 1-octene (0.4 mmol), phenylboronic ester (1.2 mmol), Pd(OAc)2 (6 mol%), Cu(OTf)2 (20 mol%), L1–L5 (20 mol%), CMP-10 (35 mg), DMA (4 mL), 40 °C, O_2 (1 atm), 48 h.

b: Linear/ branched.

c: Ratio of linear/other isomers.

5.3 Heterogenous Organocatalysis

Over the past 10 years, organocatalysis [18, 19] has played an important role in synthetic organic chemistry, and significant efforts have been made to develop organocatalysts that include both heterogeneous [20–24] and homogeneous organocatalysts. However, most homogeneous organocatalysts require high catalyst loading and are difficult to recycle from their products. These drawbacks can be overcome by the "bottom-up" construction of a robust heterogeneous organocatalyst directly from an organocatalyst monomer and a structural linker through a condensation reaction. Further advantages such as superior catalytic properties, reusability, porosity, and high surface area can also be obtained by directly embedding the organocatalyst unit in a rigid CMP framework compared to a homogeneous analog.

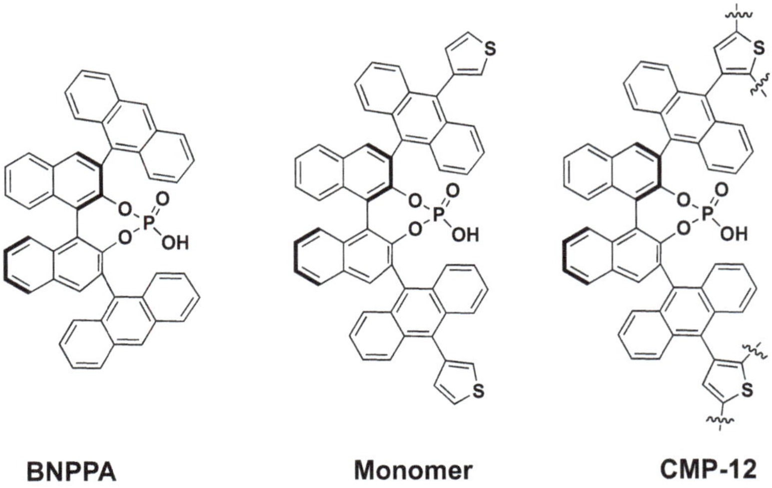

Figure 5.7 Organocatalyst used asymmetric reaction.

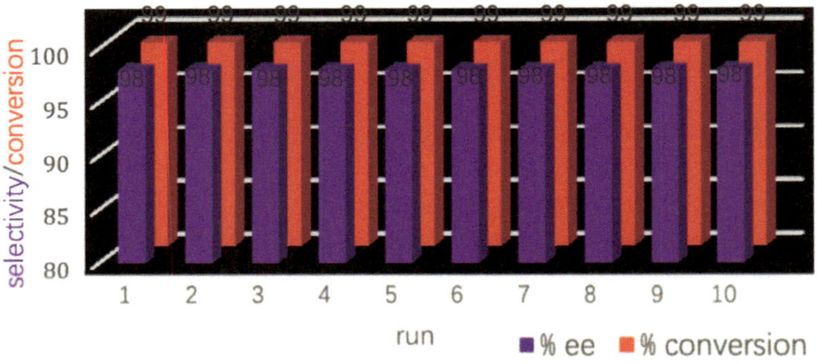

Figure 5.8 Recycling ability of CMP-12 in asymmetric transfer hydrogenation.

Thomas and Blechert's group developed an important heterogeneousasymmetricorganocatalystbyincorporatingthebinol-derived phosphate 1,1'-binaphthalene-2,2'-dihydrogenphosphate (BNPPA) into a network of porous conjugated polymers in a "bottom-up" approach [25]. The BNPPA unit introduced with a 3-thiophenyl group was polycondensed by an oxidative coupling reaction with FeCl3 to give CMP-12 (Figure 5.7). When CMP-12 was used directly as a heterologous organocatalyst, the yield was 91%, and the ee for transfer hydrogenation was 98% (Figure 5.8), which is comparable to the results for homogeneous monomers (Scheme 5.4, Formula 1). In contrast to the polystyrene-supported BNPPA reported by Rueping [26], CMP-12 requires a very short reaction time to achieve 100% conversion with the same amount of catalyst. Furthermore, the heterologous organocatalyst CMP-12 showed high activity and selectivity in asymmetric Friedel-Crafts alkylation (Scheme 5.4, Formula 2) and aza-ene-type reaction (Scheme 5.4, Formula 3) of pyrrole.

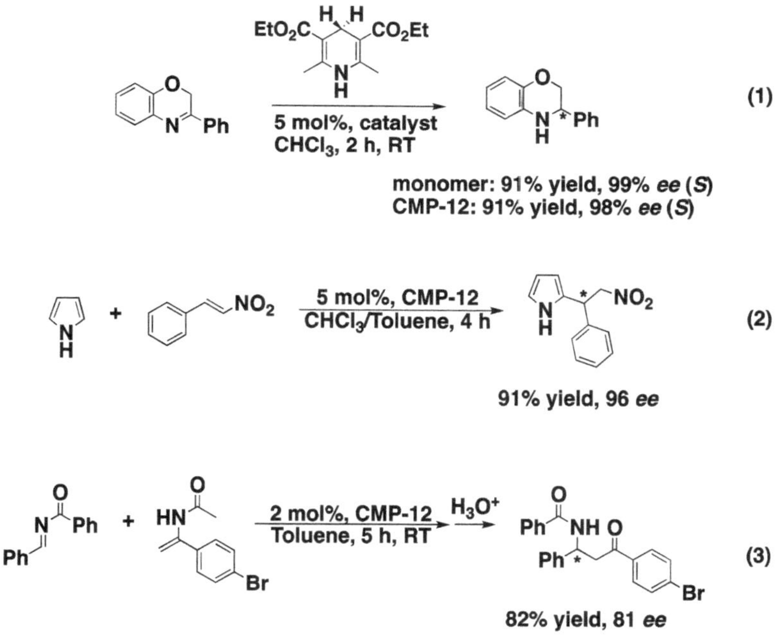

Scheme 5.4 CMP-12 used as heterogenous organocatalysts for transfer hydrogenation (Formula 1), the asymmetric Friedel-Crafts alkylation (Formula 2) and aze-ene-type reactions (Formula 3).

In another report, DMAP moieties were built in the skeleton of CMP through the Sonogashira reaction of 1,3,5-triethynylbenezene and 3,5-dibromo-N,N-dimethylpyridin-4-amine to afford CMP-13 [27]. The CMP-13 catalyst acts as an efficient heterogeneous organocatalyst for the acylation of alcohols (Figure 5.9).

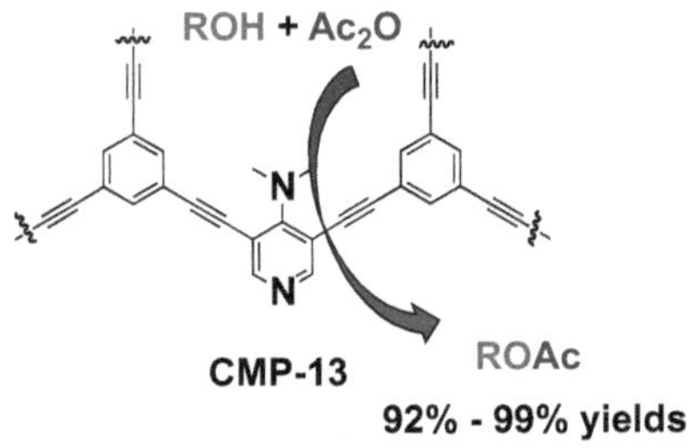

Figure 5.9 CMP-13 used as heterogenous organocatalyst for the acylation of alcohols.

A recent example presented that conjugated porphyrin framework CMP-14 was prepared via an alkyne-alkyne homocoupling polycondensation reaction, employing free-base porphyrin (5, 10, 15, and 20-tetrakis(4-ethynyl)porphyrin) as a single building block (Figure 5.10) [28]. The polymer possesses excellent catalytic substrate adaptability and good recyclability for the Knoevenagel condensation. Compared with the reported porous polybenzimidazole networks [29], the Knoevenagel condensation by CMP-14 has more rapid transformation with high yields.

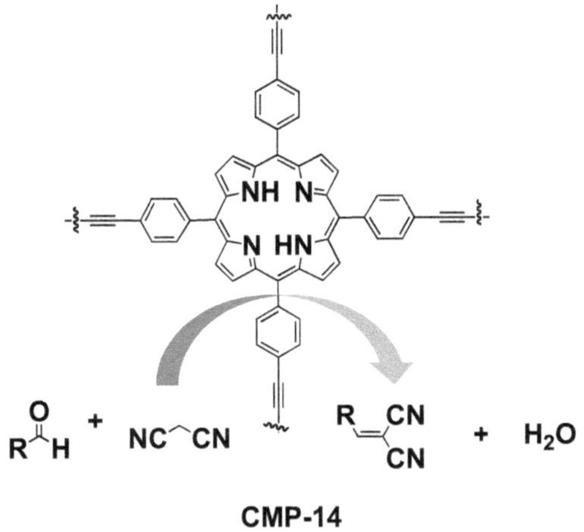

Figure 5.10 The Knoevenagel condensation of aldehydes with malonitrile catalyzed by CMP-14.

5.4 CMPs for Heterogeneous Photocatalytic Applications

A great deal of effort has been put into application. In recent years, great efforts have been made to apply CMP as a heterogeneous photocatalyst [30]. CMP has potentially diverse electronic and photophysical properties resulting from its extended p-conjugation. CMP can provide several advantages that distinguish it from other polymers in photocatalytic applications, such as flexible adjustment of absorbance and band gap attributes. CMPs also exhibit a strong light-harvesting ability that plays an important role in photocatalytic performance compared to its small-molecule analogs. Next, a series of CMP materials (Figure 5.11, CMP-15 ~ CMP-28) used as photocatalysts were designed and synthesized.

Lin et al. reported a cross-linked organic porous polymer CMP-15 containing $Ru(bpy)_3^{2+}$ chromophores in the network constructed by the Eglinton alkyne homocoupling reaction [31]. The CMP-15 was an excellent photocatalyst for the aza-Henry reaction (Table 5.3, Entry 1) and aerobic oxidative coupling of amine (Table 5.4, Entry 1) under visible light. The outstanding photocatalytic performance is attributed to excited state energy migration from the chromophores and the high content of light-absorbing $Ru(bpy)_3^{2+}$ chromophores.

Another new conjugated microporous polymer CMP-16 on BODIPY dyes developed by Liras et al., shows high activity for aza-Henry reaction (Table 5.4, Entry 2) in the presence of visible light [32]. Moreover, CMP-16 could be collected and remained efficient after seven runs.

Two new cyclometalated iridium (III) complexes (CMPs-17 a and -17 b) synthesized by Han et al. via $FeCl_3$-promoted oxidative coupling reaction can be used as highly effective and reusable heterogeneous photocatalysts in the Aza-Henry reaction (Table 5.3, Entry 3) [33].

Cooper et al. reported noble-metal-free CMP-18 synthesized by incorporating rose Bengal dye into the framework of CMP by Sonogato-Hagiwara cross-coupling polycondensation [34]. CMP-18 is highly active for photocatalytic Aza-Henry reaction and can be reused for up to 10 cycles without significant efficiency reduction

(Table 5.3, Entry 4). As an analog of CMP-18 and CMP-19, synthesized by Han et al., was also an excellent photocatalyst for the Aza-Henry reaction (Table 5.3, Entry 5) [35].

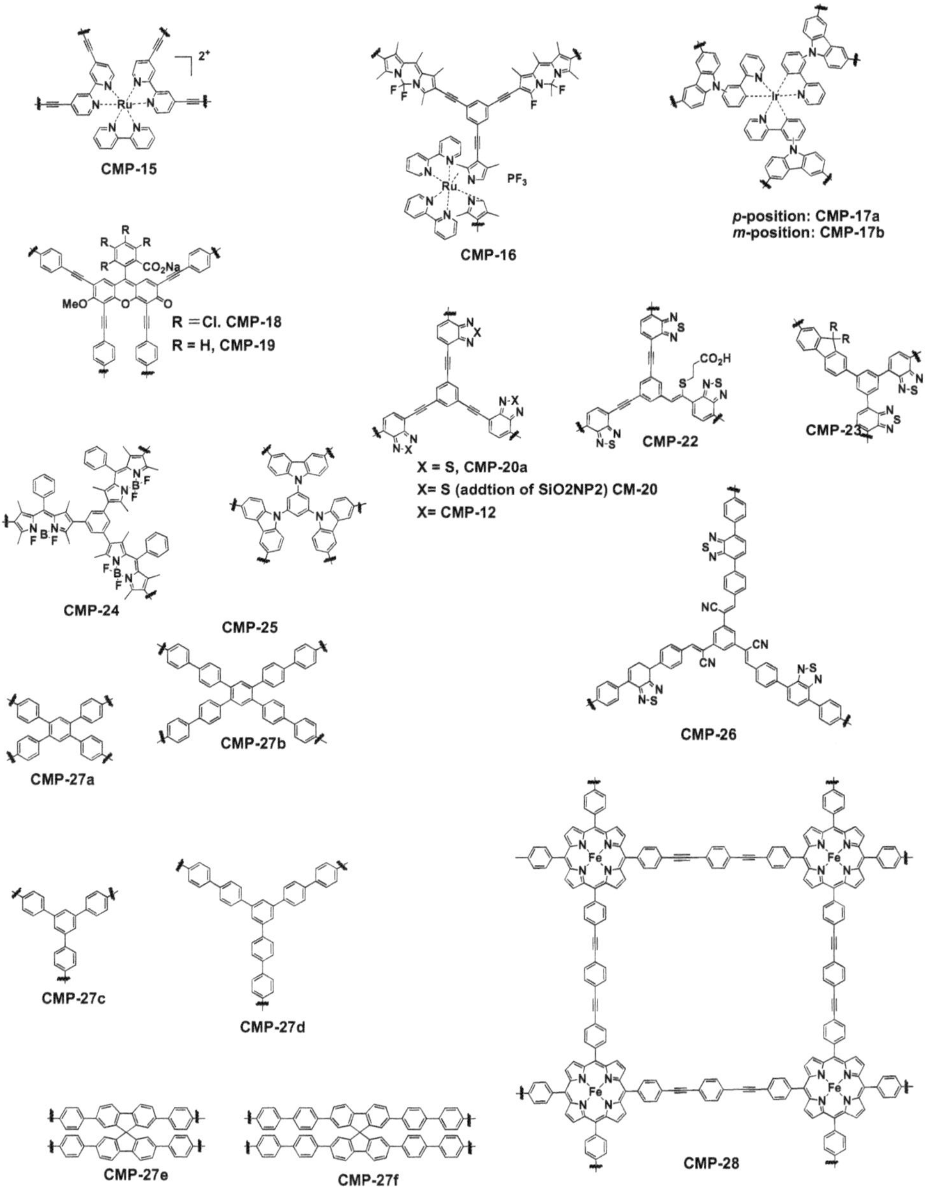

Figure 5.11 Structure of CMPs and photocatalysts.

Table 5.3 Photocatalytic aza-Henry reactions with CMPs

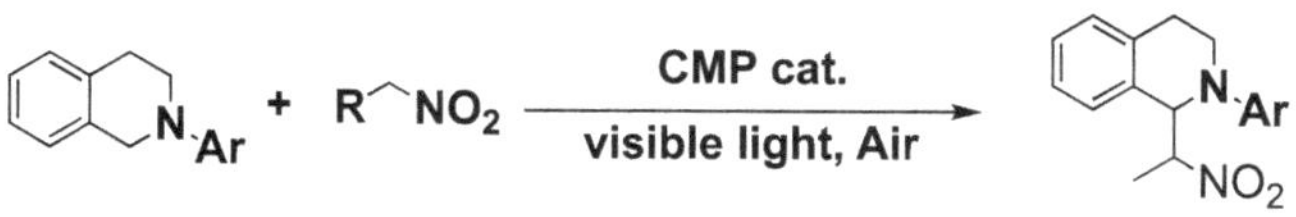

Entry	Catalyst	Light source	Yields	Ref.
1	CMP-15	Visible lamp (26 W)	85–90%	[31]
2	CMP-16	LED (15W)	82–96%	[32]
3	CMP-17a and CMP-17b	Visible lamp (23 W)	69–91%	[33]
4	CMP-18	Bulb (60W)	80–97%	[34]
5	CMP-19	Bulb (14 W)	90–98%	[35]

Table 5.4 Photocatalytic oxidative conversion of primary amines into imines

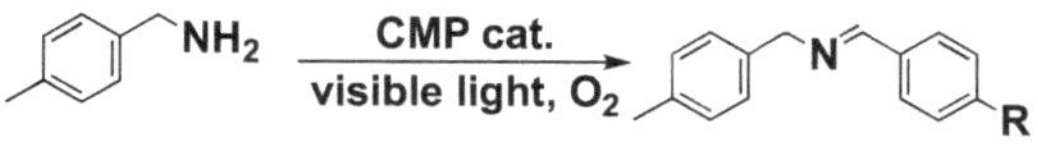

Entry	Catalyst	Light source	Yields (%)	Ref.
1	CMP-15	Xe lamp (450 W)	87–93%	[30]
2	CMP-20a	Blue LED (460 nm)	5–74%	[36]
3	CMP21	Lamp (23 W)	8–99%	[37]
4	CMP-25	Xe lamp (150 W)	94–98%	[38]

CMP-20a containing benzothiazole and CMP-21 containing benzoxadiazole were prepared by the Sinogasila-Hagiwara coupling reaction by Zhang et al. [36]. Both CMP-20a and CMP-21 exhibited excellent photocatalytic acidity for the oxidative conversion of primary amines to imines in visible light (Table 5.4). Loh et al. synthesized a novel carbazole-based conjugated microporous polymer CMP-25 by oxidative coupling polymerization [38]. CMP-25 has proven to be an active and reusable heterogeneous photocatalyst for a wide range of reactions, including selective oxidation of sulfide, aerobic dehydrogenation of inactive secondary amines, and oxidative coupling of primary amines using visible light.

CMP-20b, CMP-22, CMP-23, and CMP-24 have been found to be efficient photosensitizers capable of producing singlet oxygen (1O2) under light irradiation [39–41]. CMP-20b with benzothiazole as a building block was developed by Vilela et al. [39] by palladium-catalyzed Sonogashira coupling polycondensation. The surface area can be easily adjusted by adding SiO2 NPs during the preparation of CMP-20b. Using CMP-20b as a singlet oxygen sensitizer, α-terpinene was smoothly converted to ascaridole by Boolean light at a wavelength of 420 nm, the first example of a singlet oxygen-generating photocatalytic application of CMP (Table 5.5, Entry 1). Furthermore, CMP-23, synthesized by Vilela et al. [40], and CMP-24, synthesized by Xu et al. [41] exhibit excellent photocatalytic activity for 1O2 production (Table 5.5, Entries 2 and 3), as shown by in situ reaction with terpinene. Vilela et al. attempted to modify CMP-20 using thiol-ene chemistry to enhance water dispersibility and produced a water-heterogeneous photocatalyst for the 1O2-induced conversion of 2-furoic acid to 5-hydroxy-2(5H)-furanone using water as a solvent (Figure 5.12) [41].

Table 5.5 Synthesis of ascaridole from α-terpinene using photocatalyst

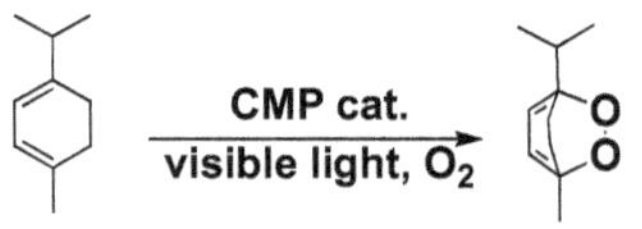

Entry	Catalyst	Light source	Conversion (%)	Ref.
1	CMP-20b	blue light (420 nm)	96%	[39]
2	CMP-23	blue light (420 nm)	100%	[40]
3	CMP-24	LED (530 nm)	100%	[41]

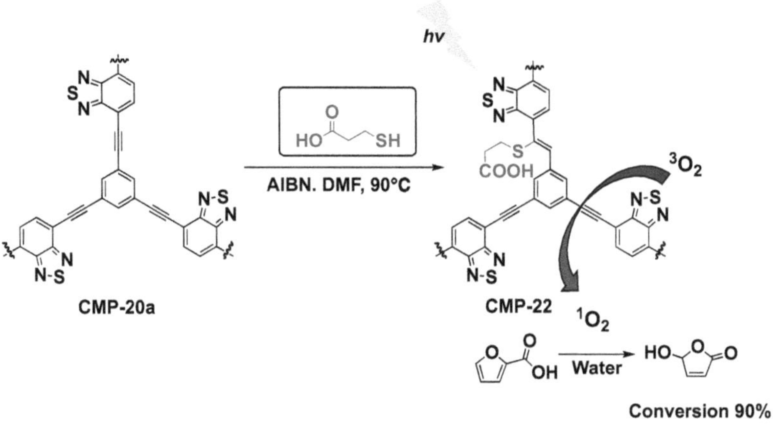

Figure 5.12 Preparation of water compatible CMP-12 and photocatalytic conversion of furoic acid into 5-hydroxy-2(5H)-furanone with CMP-22.

Cooper et al. synthesized a series of CMP materials [CMP-27 (a–f)] to investigate the effect of conjugation length and substitution pattern on photocatalytic performance for hydrogen evolution [42]. The results showed that neither the length nor the surface area of the linker correlated with the photocatalytic performance. However, a strong correlation between the photosorption profile and the photocatalytic performance has been observed.

Another photocatalyst CMP-28 with a dendrite-like nanostructure has also been successfully prepared by incorporating iron porphyrin through the Sonoto-Hagiwara coupling reaction. CMP-28 shows higher activity and degradation selectivity of Congo red than other dyes [43].

Recently, Zhang's research group reported a metal-free synthesis of CMP-26 with a high surface area via a simple Knovenagel condensation reaction under mild conditions (Scheme 5.5, Formula 1) [44]. The resultant CMP-26 was demonstrated to be a robust and efficient heterogeneous photocatalyst by oxidatively hydroxylating arylboronic acids under visible light (Scheme 5.5, Formula 2).

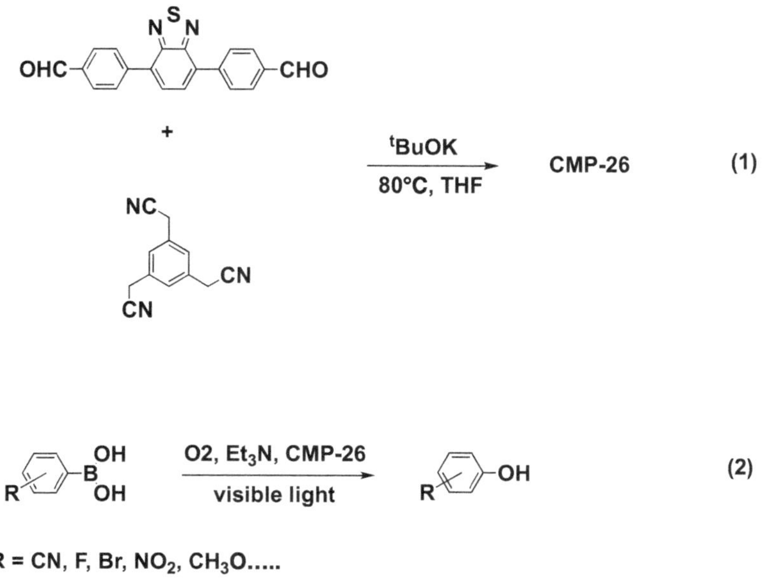

Scheme 5.5 Synthetic routes for the CMP-26 and its application in the oxidative hydroxylation of arylboronic acids.

5.5 Conclusion and Perspective

The diversity of synthesis methods and building blocks makes CMP catalysts a powerful means of modulating the pore environment and functionality. These CMP catalysts are unique in their combination of highly conjugated framework and porosity, and offer advantages in photocatalysis that other heterogeneous catalysts do not. Furthermore, due to its rigid three-dimensional structure and high stability, the CMP catalyst functions as a robust heterogeneous catalyst. Therefore, CMP catalysts have been widely developed for heterogeneous transition metal catalysis, organic catalysis, and photocatalytic applications. Many CMP catalysts are easy to isolate and recycle. The heterogeneous CMP catalyst has higher activity, stability, and selectivity than the homogeneous catalyst.

Taken together, conjugated microporous polymers are versatile platforms for functional and structural design, thus providing various opportunities for producing microporous materials for heterogeneous catalysis. However, a mechanistic explanation of the regioselectivity, stereoselectivity, and activity obtained with CMP catalysts remains a major challenge due to their amorphous frameworks. It is also difficult to remove trace metals remaining in the preparation process of the CMP catalyst. We hope these issues will be resolved in the near future.

References

1. J.-X. Jiang, C. Wang, A. Laybourn, T. Hasell, R. Clowes, Y. Z. Khimyak, J. Xiao, S. J. Higgins, D. J. Adams, A. I. Cooper, *Angew. Chem. Int. Ed.* 2011, 50, 1072–1075; *Angew. Chem.* 2011, 123, 1104–1107.
2. H. Furukawa, N. Ko, Y. B. Go, N. Aratani, S. B. Choi, E. Choi, A. O. Yazaydin, R. Q. Snurr, M. O'Keeffe, J. Kim, O. M. Yaghi, *Science* 2010, 329, 424–428.
3. B. Meunier, *Chem. Rev.* 1992, 92, 1411–1456.
4. A. M. Shultz, O. K. Farha, J. T. Hupp, S. T. Nguyen, *J. Am. Chem. Soc.* 2009, 131, 4204–4205.
5. M. J. Gunter, P. Turner, *Coord. Chem. Rev.* 1991, 108, 115–160.
6. L. Chen, Y. Yang, D. Jiang, *J. Am. Chem. Soc.* 2010, 132, 9138–9143.
7. J. Perles, M. Iglesias, M.-Á. M.-Luengo, M. Á. Monge, C. R.-Valero, N. Snejko, *Chem. Mater.* 2005, 17, 5837–5842.

8. L. Chen, Y. Yang, Z. Guo, D. Jiang, *Adv. Mater.* 2011, 23, 3149–3154.
9. L.-J. Feng, Q. Chen, J.-H. Zhu, D.-P. Liu, Y.-C. Zhao, B.-H. Han, *Polym. Chem.* 2014, 5, 3081–3088.
10. X. Wang, S.-m. Lu, J. Li, Y. Liu, C. Li, *Catal. Sci. Technol.* 2015, 5, 2585–2589.
11. J. Liu, J. M. Tobin, Z. Xu. F. Vilela, *Polym. Chem.* 2015, 6, 7251–7255.
12. H. Zhou, Y.-B. Wang, *Chem Cat Chem* 2014, 6, 2512–2516.
13. H. Zhou, Q.-Y. Zhang, X.-B. Lu, *RSC Adv.* 2016, 6, 44995–45000.
14. N. Huang, Y. Xu, D. Jiang, *Sci. Rep.* 2014, 4, 7228–7235.
15. Y.-B. Zhou, Y.-Q. Wang, L.-C. Ning, Z.-C. Ding, W.-L. Wang, C.-K. Ding, R.-H. Li, J.-J. Chen, X. Lu, Y.-J. Ding, Z.-P. Zhan, *J. Am. Chem. Soc.* 2017, 139, 3966–3969.
16. R.-H. Li, Z.-C. Ding, C.-Y. Li, J.-J. Chen, Y.-B. Zhou, X.-M. An, Y.-J. Ding, Z.-P. Zhan, *Org. Lett.* 2017, 19, 4432–4435.
17. J. H. Delcamp, A. P. Brucks, M. C. White, *J. Am. Chem. Soc.* 2008, 130, 11270–11271.
18. P. L. Dalko, L. Moisan, *Angew. Chem. Int. Ed.* 2004, 43, 5138–5175; *Angew. Chem.* 2004, 116, 5248–5286.
19. D. W. C. MacMillan, *Nature* 2008, 455, 304–308.
20. M. Benaglia, *New. J. Chem.* 2006, 30, 1525–1533.
21. M. Gruttadauria, F. Giacalne, R. Noto, *Chem. Soc. Rev.* 2008, 37, 1666–1688.
22. Y. Zhang, L. Zhao, S. S. Lee, J. Y. Ying, *Adv. Synth. Catal.* 2006, 348, 2027–2032.
23. Y. Zhang, L. Zhao, P. K. Patra, J. Y. Ying, *Adv. Synth. Catal.* 2008, 350, 662–666.
24. M. M. Parvez, N. Haraguchi, S. Itsuno, *Org. Biomol. Chem.* 2012, 10, 2870–2877.
25. D. S. Kundu, J. Schmidt, C. Bleschke, A. Thomas, S. Blechert, *Angew. Chem. Int. Ed.* 2012, 51, 5456–5459; *Angew. Chem.* 2012, 124, 5552–5555.
26. M. Rueping, E. Sugiono, A. Steck, T. Theissmann, *Adv. Synth. Catal.* 2010, 352, 281–287.
27. Y. Zhang, Y. Zhang, Y. L. Sun, X. Du, J. Y. Shi, W. D. Wang, W. Wang, *Chem. Eur. J.* 2012, 18, 6328–6334.
28. X. Liu, S. A, Y. Zhang, X. Luo, H. Xia, H. Li. Y. Mu, *RSC Adv.* 2014, 4, 6447–6453.

29. P. Makowski, J. Weber, A. Thomas, F. Goettmann, *Catal. Commun.* 2008, 10, 243–247.
30. Y.-L. Wong, J. M. Tobin, Z. Xua, F. Vilela, *J. Mater. Chem. A* 2016, 4, 18677–18686.
31. J.-L. Wang, C. Wang, K. E. deKrafft, W. Lin, *ACS Catal.* 2012, 2, 417–424.
32. M. Liras, M. Pintado-Sierra, M. Iglesiasb, F. Sánchez, *J. Mater. Chem. A* 2016, 4, 17274–17278.
33. L. Pan, M.-Y. Xu, L.-J. Feng, Q. Chen, Y.-J. He, B.-H. Han, *Polym. Chem.* 2016, 7, 2299–2307.
34. J.-X. Jiang, Y. Li, X. Wu, J. Xiao, D. J. Adams, A. I. Cooper, *Macromolecules* 2013, 46, 8779–8783.
35. C.-A. Wang, Y.-W. Li, X.-L. Cheng, J.-P. Zhang, Y.-F. Han, *RSC Adv.* 2017, 7, 408–414.
36. Z. J. Wang, K. Garth, S. Ghasimi, K. Landfester, K. A. I. Zhang, *ChemSusChem* 2015, 8, 3459–3464.
37. Z. J. Wang, S. Ghasimi, K. Landfester, K. A. I. Zhang, *Adv. Mater.* 2015, 27, 6265–6270.
38. C. Su, R. Tandiana, B. Tian, A. Sengupta, W. Tang, J. Su, K. P. Loh, *ACS Catal.* 2016, 6, 3594–3599.
39. K. Zhang, D. Kopetzki, P. H. Seeberger, M. Antonietti, F. Vilela, *Angew. Chem. Int. Ed.* 2013, 52, 1432–1436; *Angew. Chem.* 2013, 125, 1472–1476.
40. H. Urakami, K. Zhang, F. Vilela, *Chem. Commun.* 2013, 49, 2353–2355. K. Zhang, Z. Vobecka, K. Tauer, M. Antonietti, F. Vilela, *Chem. Commun.* 2013, 49, 11158–11160.
41. J. M. Tobin, J. Liu, H. Hayes, M. Demleitner, D. Ellis, V. Arrighi, Z. Xu, F. Vilela, *Polym. Chem.* 2016, 7, 6662–6670.
42. R. S. Sprick, B. Bonillo, M. Sachs, R. Clowes, J. R. Durrant, D. J. Adamsa, A. I. Cooper, *Chem. Commun.* 2016, 52, 10008–10011.
43. Z. Xiao, Y. Zhou, X. Xin, Q. Zhang, L. Zhang, R. Wang, D. Sun, *Macromol. Chem. Phys.* 2016, 217, 599–604.
44. Z. J. Wang, R. Li, K. Landfester, K. A. I. Zhang, *Polymer* 2017, 126, 291–295.

Chapter 6

CMP's Energy Storage Applications

Conjugated microporous polymers (CMPs) are emerging as an important class of materials with applications in many fields. Among these, the application in energy storage is now important and significant. The focus here is on supercapacitors, lithium-ion batteries, and sodium-ion batteries. Due to the large surface area of these materials, the unique charge storage processes found in the unique CMPs found in CMPs are discussed, and how CMPs enable ultra-fast batteries with high energy densities are reviewed. This chapter provides an overview of the methods that enable ultra-fast, high-energy-density batteries.

6.1 Introduction

Recent developments in mobile electronics, communications, and transportation systems require efficient energy storage systems with high energy and power densities [1–3]. Lithium-ion rechargeable batteries (LIB) are the most adopted energy storage systems in commercial applications, due to their excellent properties [4]. A typical configuration of a LIB is a positive electrode, a negative electrode, and a separator to avoid short circuits. The electrolyte is added to ensure the transfer of ions between the electrodes [5].

Conjugated Microporous Polymers
Atsushi Nagai

ISBN 978-981-5129-66-3 (Hardcover), 978-1-003-63652-6 (eBook)
www.jennystanford.com

Commercial anodes are usually based on graphite, and cathodes are mainly lithium metal oxides. A typical electrolyte is a lithium salt dissolved in a mixture of organic solvents. Although these materials currently provide the most powerful commercial batteries on the market, they still lack sufficient power and have significant drawbacks [6, 7]. A major drawback of commercial cathode materials based on intercalation is that their reliance on single electron transfer reactions limits the maximum achievable capacity to about 200 mAh g^{-1} under commercially feasible conditions. In addition, the materials used for the electrodes are mostly rare earth elements with a finite supply. Such problems will continue to hinder price reductions for such batteries, which are still very high compared to alternative non-renewable sources such as gasoline [3, 8, 9]. The production and manufacture of batteries generates a variety of pollutants and hazardous substances, including toxic fluoride gas, various metal and carbon dusts, and solid waste, and causes serious environmental problems, including water, land, and air pollution, and workplace hazards [10, 11]. Safety is another major concern, which includes equipment safety and worker safety [12]. Workers exposed to hazardous substances during processing can develop serious health problems, and the use of flammable and volatile substances in equipment poses a serious fire hazard. This scenario indicates a high need to develop sustainable and alternative energy storage systems, which requires finding new materials or developing new battery chemicals [1, 13–17].

Over the past few decades, organic materials have emerged as promising candidates for next-generation lithium-ion batteries and supercapacitors [5, 18, 19]. They offer several advantages over inorganic electrode materials: (a) organic materials are usually inexpensive and abundant in nature; (b) electrochemical properties such as redox potential or theoretical charge storage capacity can be easily tuned by adjusting their structure, which is much more difficult in the case of inorganic materials; and (c) organic electrode materials are versatile in that if they are electrochemically active for lithium ions, they are almost equally active for other metal ions such as sodium and magnesium, paving the way for using the same material in different battery chemistries [20–22]. In addition, most inorganic electrode materials are capable of only single electron transfer reactions; whereas, organic materials typically exhibit multiple electron transfer reactions [23]. Since the energy storage mechanism in redox reactions is based on conversion reactions, such multiple electron transfer processes do not contribute to the structural degradation of the electrode (Conversely, in the case of inorganic materials, multiple electron transfer processes lead to severe structural degradation in intercalation-based mechanisms), thus ensuring the long-term stability of the electrode. Therefore, this reaction mechanism can contribute to the construction of powerful batteries with high energy density based on organic electrode materials. Such organic materials have several advantages but also many disadvantages [24]. For example, most redox-active materials have low electrical conductivity. Small organic molecules usually exhibit high theoretical capacities but low-rate performance and stability due to their solubility in liquid electrolyte systems. Therefore, the practical energy density and experimental capacity of organic materials remain low due to the low utilization of active materials.

To remedy these challenges, several strategies have been developed, including composite formation and flexible binder-free electrode films [25–28]. Polymerization is an effective strategy to introduce higher stability but often reduces the theoretical specific capacity [24]. This decrease in specific capacity is caused by entanglement and stacking that make it difficult for the electrolyte to access the active sites, which, together with the low conductivity, leads to a decrease in the utilization of the active material. Thus, a major challenge is to design novel organic materials with sufficient capacity, good conductivity, and accessible ion diffusion pathways to maximize active material utilization in next-generation energy storage devices. Porous organic polymers are expected to provide solutions to many of the challenges described in this field [29]. Figure 6.1 schematizes some of the problems associated with small organic molecules and why porous molecules are being considered as more promising candidates.

Porous materials are the materials of choice for many applications due to their attractive properties compared to non-porous materials, such as high surface area that can provide more active sites for chemical reactions and channels (connected or not) that may make buried active sites accessible. Within the versatile family of porous materials, conjugated microporous polymers (CMPs) have emerged as an important material class due to their wide range of applications in the broad areas of energy and environment [30]. CMPs are defined as polymers with π-conjugated backbone, composed of purely covalently bonded organic materials with pore sizes less than 2 nm with a pore size of less than 2 nm [31]. Compared to other organic and inorganic materials, CMPs have a variety of attractive properties including extended π-conjugated structure, uniform micropores, very good chemical and thermal stability, ease of functionalization with various organic groups, high surface area, and adjustable properties [30–35]. In addition to these features, CMPs can synthesize a variety of structures, including two- and three-dimensional networks, dendrimers, and branched polymers [33]. An important aspect of CMP is the precise control of pore size, chemistry, and overall structure. Because of these attractive properties, CMP has been applied in numerous applications, including catalysis, gas storage, light harvesting, sensing, luminescence, and other optical and energy storage applications [36–44].

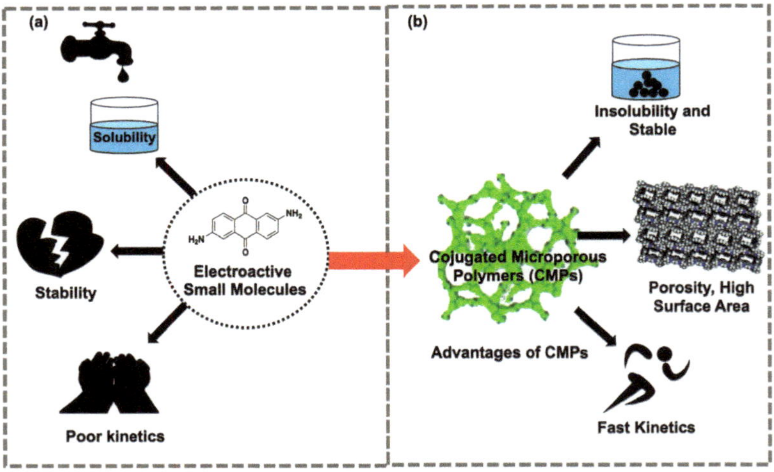

Figure 6.1 (a) Challenges associated with electroactive small organic molecules and (b) some of the advantages of organic conjugated microporous polymers.

The scarcity of lithium resources, combined with sustainable device production, and the bias in their geographical distribution is another important issue that forces the exploration of alternative battery chemistry. Researchers have begun to study various metals such as sodium, copper, zinc, potassium, and magnesium as alternatives to lithium [45–48]. Among these candidates, sodium appears to be the most promising candidate because of its natural abundance, worldwide availability, and suitable redox potential.

Recently, a number of reviews have been published summarizing advancements in the field of CMP and COF [31, 40, 49]. There are also reviews discussing the application of porous polymers and COFs to batteries [50, 51]. However, they are too limited to only specific classes such as COFs, and do not discuss the challenges and problems associated with CMP in the energy storage sector. Articles that not only summarize the latest advances but also highlight problems and provide empirical insights to solve those problems are therefore the call of the age. In this light, the latest advances in the field of CMP for energy storage applications are summarized and some prominent studies are taken as examples.

It, especially, focuses on lithium-ion batteries and sodium-ion batteries. It should be noted that there is already a large body of literature on the design, synthesis, and classification of CMPs, so the focus here is only on the application part [31, 49]. It also provided meaningful insights into the development of ultrafast rechargeable batteries that take advantage of the improved porosity and conductivity of the optimized surface area of CMP. This consideration is useful for the material design of high-power and high-energy density batteries. It further discusses the main challenges facing CMPs in energy storage and highlights the points that need further investigation to realize the potential of CMPs to contribute to energy storage. It stresses that further investigations are needed to realize the potential of CMP to make a significant contribution to emerging battery technologies.

6.2 CMPs as Electrode for LIBs

Redox-active polymers have been used as electrode materials in batteries for decades [52]. Han et al., in 2007, used a carbonyl derivative conjugated polymer as a positive electrode for LIBs, which was the first report of conjugated polymers in energy storage [53]. In particular, the paper by Jiang et al. appears to be the first report on the use of CMP in energy storage [54].

In recent years, several new studies have been initiated on the energy storage properties of CMPs, especially in LIBs and SIBs. The next section summarizes progress in this area. The availability of molecular design parameters and many organic redox-active functional groups allows careful tuning of the structure and redox properties of CMPs. Based on the redox-active groups present, CMP can be used as a positive or negative electrode. Similarly, if CMP is used in LIBs, it is likely to perform well in sodium-ion and potassium-ion batteries. This chapter focuses on lithium-ion and sodium-ion batteries.

6.2.1 Comparison of Covalent Organic Frameworks (COFs) and CMPs

Although the focus of this book is on CMP, it is important to highlight the differences and similarities between CMP and COF, as both CMP and COF are subclasses of porous organic polymers (POPs). CMP and COF are two- or three-dimensional structures formed by carbon, hydrogen, and a light framework formed by heteroatoms such as nitrogen, sulfur, boron, and oxygen. Both can be synthesized by various reactions, but to our knowledge, most reported COFs are produced by condensation reactions such as Schiff base chemistry; whereas, CMP is mostly synthesized by cross-linking reactions such as Buchwald-Hartwig [55–57]. The main difference between these subclasses of POPs is that COFs are crystalline while CMPs are amorphous, which results in different properties. Since COF is crystalline, it is easier to characterize and process than CMP, but its thermal and chemical stability is inferior. The main property of CMPs is an intrinsically advanced conjugate structure that is not required for COFs. However, COFs are like CMPs because they can be microporous as well as conjugated. In this respect, CMP can be regarded as an amorphous version of conjugated COF [31]. With these differences and similarities in mind, I discuss conjugate COFs with CMP. In recent years, electroactive COFs have been much studied in the area of energy storage because of their interesting properties, and a comprehensive progress report can be found elsewhere [50, 58, 51, 29].

6.2.2 CMPs as Cathode Materials for LIBs

High energy density batteries require cathode materials with high specific capacity and high redox potential. Although there are many reports on organic cathode materials with high specific capacity, it is very difficult to develop redox-active organic cathode materials with a redox potential of more than 2.5 V vs. Li/Li^+. In an interesting report, researchers developed a triphenylamine (TPA)-based microporous polymer that showed a discharge plateau at 3.6 V vs Li/Li^+ [59] (see Figure 6.2a and b). An electron transfer redox reaction between the neutral and radical states of TPA was proposed as a possible charge/discharge mechanism. The team concludes that redox-active units organized in a microporous framework with a high surface area can help realize batteries with high power and energy densities.

Reducing the inert materials and increasing the content of redox-active units in the polymer structure is a very effective approach to synthesizing redox-active polymers with theoretically high energy storage capacity. This strategy has recently been reported by researchers to have the highest practical capacity of 502.4 mA g^{-1} [60]. The researchers also recorded a capacity retention of 81% after 1000 charge/discharge cycles at a current density of 1.54 A g^{-1}. These results indicate that increasing the content of redox-active units in the π-conjugated stable porous framework not only increases capacity but also effectively improves stability.

A p-type conjugated quinoxaline-based heteroaromatic molecule (3 Q) was reported for high power density organic LIBs [61]. A composite electrode using 60% graphene (w/w) as a conductive additive was assembled and its electrochemical performance was analyzed. In an interesting study, hexahydroxytriphenylene was condensed with 1,4,5,8,9,12-hexaazatriphenylene -2,3,6,7,10,11-hexacarbonitrile to obtain a crystalline but porous charge/discharge composite named PCT-1 [62]. Hexagonal prismatic porosity is mixed. When PCT-1 was used as the positive electrode, it showed a capacity of 288 mAh g^{-1} at 500 mA g^{-1}. The reported cycling performance of the composite also contributes to the stabilization of redox molecules in the liquid electrolyte. However, a drawback of this study is that the active material content remains at 30% (w/w), which is very low compared to commercial applications where the active material typically accounts for 85% of the composite electrode.

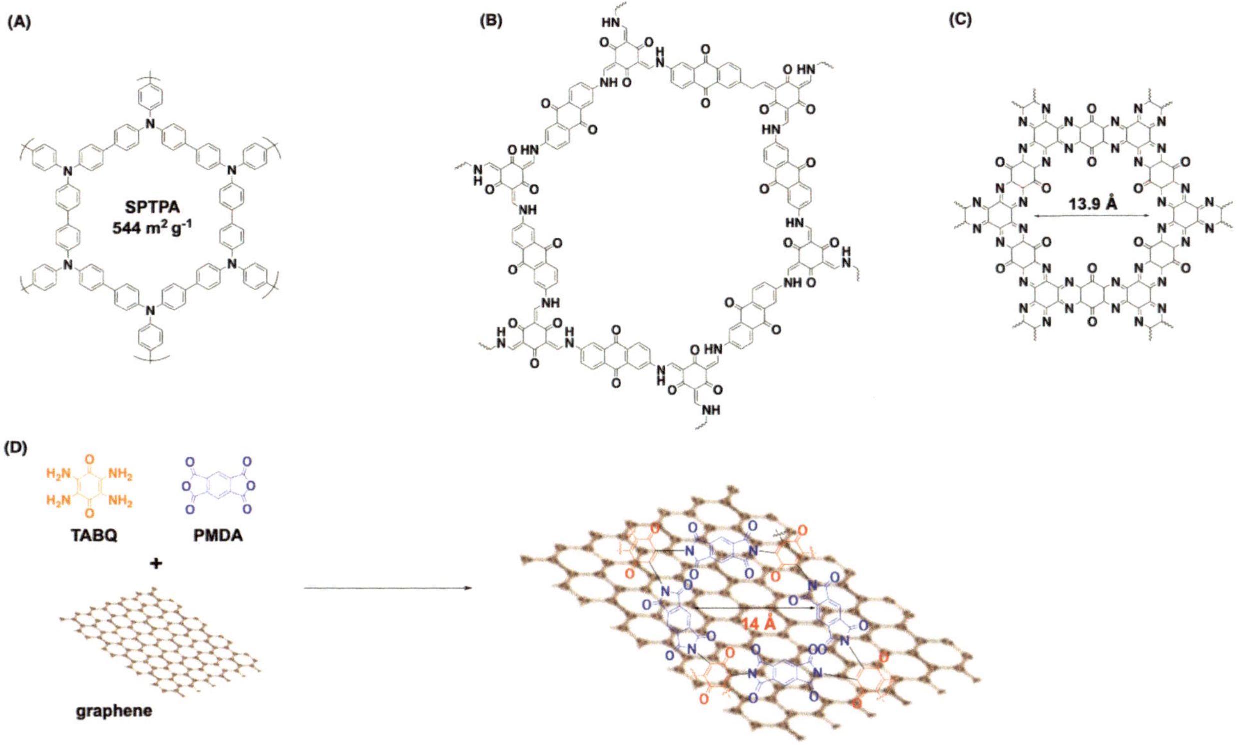

Figure 6.2 Few representative examples of CMPs being used as cathode materials, (A) SPTPA cathode, (B) and (C) nitrogen-rich 2D graphitic framework, and (D) carbonyl-rich porous organic framework.

Conjugated microporous polymers of 1,3,5-tri (4-diphenylaminophenyl) benzene (TTPAB) have been synthesized with a spherical morphology [63]. The surface area was 595 m^2 g^{-1} and both microporous and mesoporous structures were recorded. For comparison, another CMP polytriphenylamine (PTPAn) was also synthesized. Both materials (PTTPAB and PTPAn) were found to have different morphologies. When used as a positive electrode in lithium-ion batteries, TTPAB showed a much better rate and cycling performance under similar conditions than PTPAn. Electrochemical impedance spectroscopy (EIS) studies revealed that the electrode dynamics of TTPAB were superior to those of PTPAn, and the authors concluded that in the case of PTPAn, the compact structure and plate morphology prevented the fast diffusion of ions; whereas, in the case of TTPAB, the porous morphology facilitated the diffusion, leading to better electrochemical properties. This study shows that the morphology of CMP contributes to the determination of charge storage properties. The amounts of conductive additives used were relatively high (40% of the total weight of the electrode materials).

The synthesis of COFs with high specific capacity is difficult, but more research has been conducted recently. A polyimide benzoquinone (PIBN)-based COF with a high theoretical specific capacity was prepared due to its abundance of redox-active groups [64]. The poor electrochemical performance of this material was improved by compositing it with graphene (PIBN-G). When this was applied to the cathode of a lithium-ion battery, a high specific capacity of 271 mAh g^{-1} was obtained, and the active material utilization ratio reached 96%. The positive electrode also exhibited satisfactory performance with good rate performance and cycle stability. The strong interaction between graphene and PIBN, the easy charge/discharge process due to the improved conductivity, and the ion diffusion due to the porous nature were found to be the main factors for the improvement of these properties. In an interesting report, naphthalene diimide (NDI) was used to prepare thin films of β-ketoenamine-linked porous polymer networks (NDI-TFP) containing NDI [65]. Cations from the supporting electrolyte were found to stabilize the reduction process of NDI-TFP thin films. Because of the porous nature of NDI-TFP, cations could be retained in the pores and stabilize the dimer structure. In the case of NDI-TFA polymer films, a positive shift in redox potential was observed compared to the NDI dimer. The author concluded that microstructure is also an important parameter to consider when designing energy storage devices.

Recently, there has been interest in redox-active β-ketoenamine-bonded COFs for energy storage applications due to their high surface area and presence of redox-active sites (Figure 6.2c). However, these electrochemical sites are not fully accessible due to their rigid and non-conductive structure. One way to improve performance is to use a mechanical exfoliation method to fabricate multi-layer thick nanosheets of β-ketoenamine-bonded COFs [66]. When these nanosheets were applied as positive electrodes in LIBs, the rate performance and cycling performance were improved compared to unexfoliated COFs. Careful molecular design improved the discharge voltage to 3.6 V vs Li/Li^+ and the theoretical capacity to 210 mA g^{-1}. As evidenced by this study, the combination of exfoliation strategy and molecular design could be a useful approach to improve the energy and power densities of LIBs. Electro-polymerization of the conductive polymer PEDOT within the pores of the COF further improved the electrochemical performance of the COF [67]. Although this composite was not directly used in a battery, it was shown that the conductivity and redox properties of this COF were significantly improved and the utilization of active materials was improved. In a related report, PEDOT was polymerized within nanochannels of anthraquinone-linked COF (AQ COF) [68]. In the final composite, PEDOT acted as a highway for electron transport, resulting in high electrical conductivity (synthesized with the parent AQ COF). Although PEDOT functionalization was an excellent strategy to improve the electrochemical properties of COFs, it had the disadvantage of reducing the overall bulk density due to the addition of an extra inert material. To circumvent this problem, the same group developed a new phenazine-based 2D COF (DAPH-TFA COF). In a recent report, Wei et al. reported an AQ-based COF on CNTs by an in situ method, showing excellent rate and cycling performance [70]. Here, the combination of high conductivity and large porosity contributes to the attractive performance of these composites.

Porous graphene-like COFs (sometimes called graphite frameworks) have been studied for LIBs (Figure 6.2d). These nitrogen-doped graphite frameworks have a redox potential of about 2.5 V and are suitable as cathode materials. Hexaazatrinaphthalenebased nitrogen-rich 2D COF materials have been reported for application in high-performance batteries [71]. Recent reports have found that the use of basic conditions during synthesis improves the crystallinity and electrochemical redox activity of sacred graphene-like COFs [72–74]. To achieve high theoretical capacity, a carbonyl-rich microporous framework was grown on graphene and applied as a cathode [64]. Normally, the theoretical capacity of COF decreases due to the presence of a core without redox activity. An interesting report on benzoquinone-based COF (BQ COF) is that they achieve a high theoretical capacity of 517 mA g^{-1} with relatively large redox-active groups per molecule [60].

6.2.3 CMPs as Anode Materials for LIBs

The oxidation-reduction potential of the material used as the negative electrode is preferably close to 0 V if not exactly 0 V (vs. the lithium electrode for LIB). The potential of a commercially available graphite negative electrode is generally 0~0.25 V, which is suitable as a material used as a negative electrode but the theoretical capacity is limited to 371 mAh g^{-1}. A 4,7-dicarbazyl-[2,1,3]-benzothiadiazole (PDCCzBT)-based anode material was reported for lithium- and sodium-ion batteries [75]. Two processes, doping and de-doping of lithium ions into polymer chains and surface adsorption of lithium ions, were found to be responsible for energy storage in this material. A capacity of 1042 mAh g^{-1} was demonstrated due to the high specific surface area of the polymer when applied as a negative electrode in lithium-ion batteries. The authors suggested that the unique structure and homogeneous microporosity of PDCzBT are the main reasons for the relatively good rate and cycle performance in both lithium- and sodium-ion batteries.

Thiophene-rich CMPs were constructed using a 4, 8 di (thiophen-2-yl) benzo[1, 2 b: 4, 5 b'] dithiophene unit and used as negative electrodes in lithium- and sodium-ion batteries [76]. It is suggested that the fully conjugated framework supporting the charge transport property, the low band gap of the thiophene unit, the high redox activity, and the stable structure lead to very good charge storage properties in both LIBs and SIBs. This study shows that adjusting the degree of cross-linking or increasing the number of redox-active thiophene units per molecule can effectively enhance the electrochemical energy storage properties of materials.

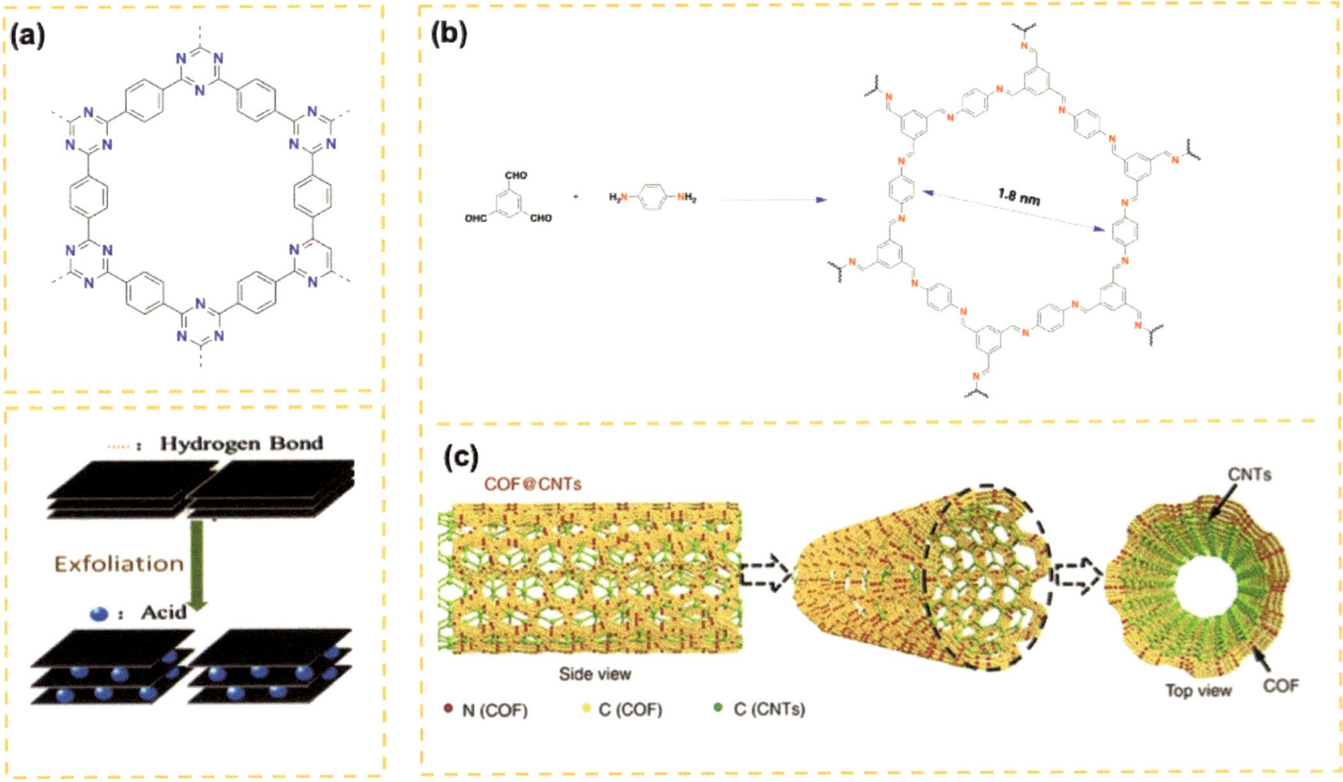

Figure 6.3 Selected examples for CMPs as anode: (a) a covalent triazine framework was exfoliated after synthesis to achieve high performance, (b) and (c) A COF was grown on CNTs to improve the conductivity and charge transfer processes. (a) Reproduced from Ref. [77] with permission from The Royal Society of Chemistry; (c) Reproduced from Ref. [78] available under a Creative Commons (CC). License by Springer Nature.

Recently, a layered covalent triazine framework (CTF) was reported as a negative electrode for LIBs (Figure 6.3a) [77]. An acid-dependent intercalation method was used to exfoliate CTF into single or multi-layered nanosheets. The ion diffusion was accelerated by the peeling, and the specific capacitance increased by more than 2% in the peeled sheet. To improve the electronic conductivity of COF, 1,3, 5-venezene tricarbaldehyde and 1,4-diaminobenzene were condensed with CNTs to obtain a COF@CNT structure with several COF layers on the CNT surface (Figure 6.3b and c) [78]. When COF was grown on CNTs, the benzene ring was found to be activated for the energy storage process. When these materials were applied as negative electrodes, a lithium-ion storage capacity of 1536 mAh g^{-1} at 100 mA g^{-1} was recorded. This is one of the highest lithium-ion storage capacities to date for which COF has been applied as a negative electrode. In the long-term cycle performance test, no capacity degradation was observed after 500 cycles of charge and discharge at 100 mA g^{-1}.

From this consideration, CMP can be effectively used as both the negative and positive electrodes of LIBs. Table 6.1 outlines the important parameters. Clearly, in order to develop a commercial need, i.e., a high-density battery, it is necessary not only to improve the conductivity of the active material to reduce dead mass but also to pay attention to the working potential.

Table 6.1 A comparative analysis of different features of CMPs in LIBs application

CMPs	Surface area (m^2 g^{-1})	Working potential (V)	Highest capacity (mAh g^{-1})	Current density (mA g^{-1})	Active loading (% age)	Ref.
YPTPA	1557	2.5–4.2	105.7	200 (1.82 C)	30	[59]
BQ COF	94.73	1.2–3.5	502.4	38.65 (0.05 C)	50	[60]
3Q	-	1.2–3.9	395	400 (1 C)	30	[61]
PTTPAB	595	2.5–4.2	84	50	50	[63]
PTPAn	11	2.5–4.2	82	50	50	[63]

(*Continued*)

Table 6.1 (*Continued*)

CMPs	Surface area ($m^2 g^{-1}$)	Working potential (V)	Highest capacity ($mAh g^{-1}$)	Current density ($mA g^{-1}$)	Active loading (% age)	Ref.
PIBN-G	-	1.5–3.5	271	28 (0.1 C)	80	[64]
DAAQ-ECOF	216	1.5–4.0	145	20	60	[66]
DABQ-ECOF	-	1.5–4.0	210	20	60	[66]
DAPH-TFP COF	1155	1.4–3.6	81.7	85.5 (0.5 C)	60	[69]
PGF-1	101	1–3.6	842	100	-	[71]
2D CCP-HATN	317	1.2–3.9	90	100	80	[72]
PDCzBT	1166	0.1–2.9	1042	20	60	[75]
DBD-CMP2	38	0.005–3.0	995	100	50	[76]
f-CTF-1	-	0.05–3.0	816	100	70	[77]
COF@CNTs	44	0.005–3.0	1536	100	80	[78]

6.3 CMPs as Electrode for SIBs

Rechargeable SIBs have emerged as alternative electrochemical energy storage systems because sodium resources are rich in raw materials, safe, and low cost. Due to the similar physicochemical properties of sodium and lithium, the energy storage chemistry and electrode material selection of organic SIBs are like those of LIBs [2]. SIBs suffer from the drawback that due to the larger ionic radius of Na^+ (102 pm) compared to Li^+ (76 pm), strain is increased during the insertion/extraction process, which eventually leads to the collapse of the rigid 3D network of inorganic electrode materials [79]. Because organic electrode materials store energy by redox reactions rather than by intercalation (as inorganic materials do), large cation sizes do not cause structural damage. Organic electrode materials with sufficient redox-active sites have the potential to accommodate large amounts of Na+ ions, but, as mentioned above, have disadvantages such as dissolution of the active material in the electrolyte and low conductivity. Therefore, a major challenge in improving the performance and lifetime of SIBs is the development of highly adaptable electrode materials [80].

Electroactive CMPs are also being investigated in sodium-ion batteries (SIB) due to their attractive and tunable properties [75, 81–83]. Due to the energy storage mechanism, large surface area, and microporous structure, CMP can easily accommodate a large volume of Na^+ ions without experiencing volume expansion or strain [84, 85]. In addition, it is easy to achieve high theoretical capacity in CMP by designing materials with abundant redox-active groups. Since CMP has a high crosslinking property, solubility in an electrolyte solution is low. Hereinafter, the performance of CMP as an electrode in SIB will be described.

6.3.1 CMPs as Cathode Materials for SIBs

The most attractive property of a material used as a positive electrode is its high redox potential. Because the standard electrode potential of lithium is more negative than that of sodium, the SIB cell voltage is generally lower [86]. Thus, finding a cathode material with a sufficient positive electrode potential (vs. sodium) or high practical capacity to compensate for this effect is a challenging task. Sakaushi et al. reported a microporous COF-based cathode material consisting of an active triazene ring and an inactive benzene ring for SIBs, showing specific capacities of approximately 200 and 10 mA g^{-1} at 0.01 and 10 A g^{-1}, respectively [87]. Reported COFs suffer from low discharge capacities on the order of 200 mAh g^{-1} and poor rate performance. This limits their further application in energy storage systems [88, 89]. It was found that the introduction of N atoms into the CMP structure efficiently participates in the redox reaction of Na-ions and finally increases the theoretical capacity. The addition of N atoms causes a decrease in the energy gap between the lowest unoccupied molecular orbital (LUMO) and the highest occupied molecular orbital (HOMO), resulting in an increase in the electronic conductivity ($\sim 10^{-9}$ S cm^{-1}) and an increase in the ionic conductivity ($\sim 10^{-4}$ S cm^{-1}) [81, 90, 91]. Chen and co-workers reported TQBQ-COF with honeycomb-like morphology with triquinoxalinylene and benzoquinone units with dual redox active sites and nanoporous structure for high energy and high-power SIBs (see Figure 6.4a for structure and mechanism). The TQBQ-COF cathode showed a high reversible capacity of 452 mAh g^{-1} and maintained 352.3 mAh g^{-1} after 100 cycles at a current density of 0.02 A g^{-1} in SIB. Furthermore,

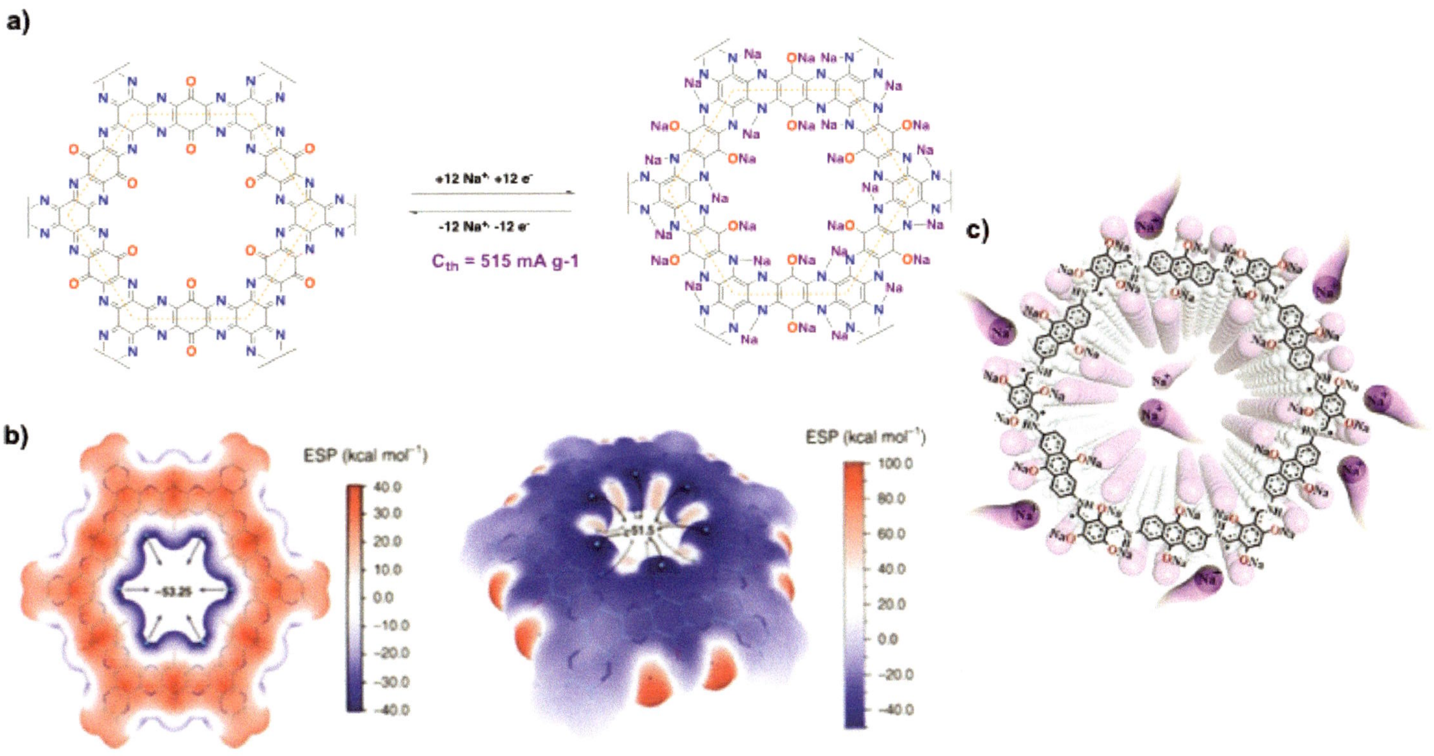

Figure 6.4 Application examples of CMP: (a) proposal of chemical structure and electrochemical redox mechanism of graphite skeleton as cathode, (b) electrogram sodiation and desordiation by electrostatic potential (MESP) method, and (c) anthraquinone-based CMP as anode of SIB. (b) Ref. [81] is available under Springer Nature's Creative Commons (CC) license; (c) [138] copyright 2019, American Chemical Society, reprinted with permission.

these insoluble COFs exhibited excellent cycling stability, with a capacity retention of 96% after 100 cycles at 1.0 A g^{-1}. Interestingly, their results showed that both carbonyl and pyrazine moieties could react with Na^+ ions simultaneously during the sodiation/desordiation process (corresponding to 12 Na^+ ions per 1 COF unit). Molecular electrostatic potential maps for sodiation and desodiation are shown in Figure 6.4b [81].

Similarly, S doping in carbonaceous materials has been shown to improve sodium-ion storage, capacity, and electronic conductivity by introducing defects and pores and increasing interlayer distances [92, 93]. Researchers have also reported a highly porous azo-bonded polymer (ALP) as a new redox-active electrode material for SIBs for rechargeable batteries [94]. In their study, ALP showed an average specific capacity of 170 mAh g^{-1} at a rate of 0.3 °C and electrolytic efficiency of 96%. It was found that the high surface area and p-conjugated microporous nature of ALP are mainly responsible for the adsorption of the electrolyte into the pores, resulting in the enhancement of ion transport and charge transfer. The pentacene tetraone system was employed to construct a highly porous p-conjugated system [95]. The excellent rate and cycling performance they observed are attributed to the fact that the extended p-conjugated system stabilizes the framework and improves charge transport.

6.3.2 CMPs as Anode Materials for SIBs

Electroactive CMP has also been studied as an anode for SIBs because of its low redox potential. 4,7-dicarbenzyl-[2,1,3]-benzothiadiazole (PDCzBT) was reported as a negative electrode material for SIBs because of its high surface area and abundant redox active sites [75]. At a current density of 29 mA g^{-1}, a reversible capacity of 145 mAh g^{-1} was obtained after 100 cycles, and a high electrolytic efficiency of more than 96% was obtained after 15 cycles. Researchers have found that the conductivity of COFs with poor conductivity can be improved by carbonization, resulting in unique porous carbon materials with improved electrochemical performance as negative electrodes for SIBs [96–98]. In 2017, Zhang et al. reported N-doped porous carbon materials obtained from carbonization of triformylphloroglucinol (TP)-*p*-phenylenediamine (PA) (denoted as TPPA) COF at various temperatures and employed the resulting nanoporous carbon (NPC) as a negative electrode material for sodium-ion batteries [99]. This material showed good rate performance and cycle performance. This type of NPC is characterized by a porous structure with N-doping, high specific surface area, and high interlayer stability.

An interesting study, recently reported by Lu and co-workers, revealed the formation of metastable radical intermediates during the electrochemical charge/discharge process (Figure 6.4c) [100]. They prepared *β*-ketoenamine-linked DAAQ-COF by a simple condensation reaction. The authors proposed that interlayer stacking interactions between adjacent COF layers play an important role in enhancing electrochemical performance and cycling stability. It is proposed that it plays an important role in improving the electrochemical performance and cycling stability of two-dimensional COF. When this COF was used as the anode of SIBs, excellent rate and cycling performance were obtained. This result indicates that the self-exchange behavior of interlayer electrons is suppressed by decreasing the stacking thickness, and more redox active sites become available. This approach also improves the stabilization of radical intermediates, ultimately leading to high-rate performance and cycle stability. In 2018, Kim et al. investigated the Na insertion properties of various network polymers (covalent organic nanosheets, CON) as anode materials for Na batteries [101]. This study demonstrated that cross-coupling of two monomers under reflux or solvothermal conditions can result in various CONs with variable porosity, network connectivity, and conductivity, showing that the specific surface area, self-assembled morphology, and molecular planarity can be effectively controlled by careful choice of monomer combination and synthetic route. As the Na-ion storage capacity of nanosheets mainly depends on the CONs' frameworks and morphology, significant enhancement of their discharge capacities was possible by tuning the structure of the framework. Among all the synthesized CONs, CON-16 delivered the best cyclic performance, rate capability, and reversible discharge capacity of 250 mAh g^{-1} after 30 cycles at 100 mA g^{-1}.

CMP with adjustable porosity and heteroatom doping level is an ideal precursor for in situ preparation of heteroatom-doped carbon materials. For example, Li et al. prepared S-doped carbon materials by thermal decomposition of CMP and showed excellent

Na-ion storage performance [102]. Furthermore, it was found that sulfur can significantly change the structure of carbon materials at high temperatures by creating defects, disorder, pores, and widening interlayer spacing, increasing electronic conductivity, enhancing the diffusion of Na-ions in battery systems, and enhancing electrochemical utilization. When applied as a negative electrode of SIB, a discharge capacity of 440 mAh g^{-1} at 50 mA g^{-1}, excellent rate performance, and excellent cycling performance of 297 mAh g^{-1} at 500 mA g^{-1} after 1000 cycles were achieved. This performance is attributed to the extension of the lattice distance of the sulfur-doped carbon after S doping.

Table 6.2 A comparative analysis of different features of CMPs in the SIBs application

CMPs	BET (m^2 g^{-1})	Working potential (V)	Highest capacity (mAh g^{-1})	Current density (mA g^{-1})	Active material loading (% age)	Ref.
BPOE	-	1.3–4.1	200	10	70	[87]
TQBQ-COF	94.36	1.00–3.0	452	20	50	[81]
ALPs	550	0.01–3.0	170	80 (0.3 C)	60	[94]
DAPT-TFP-CPF	68.02	0.8–3.2	145	00	50	[95]
PDCzBT	1166	0.10–0.9	145	20	60	[75]
NPC-2	482.06	0.005–3.0	237.7	40	-	[99]
DAAQ=COF	-	0.05–3.0	420	100	60	[100]
CON-16	300	0.01–2.5	250	100	70	[101]
SCs	331.4	0.005–3.0	440	50	80	[101]

These early reports show very promising results for CMP as an electrode material for sodium-ion batteries. However, several challenges still exist in this area and more intensive efforts are needed to solve these problems. For example, as noted above, the cell voltage of the SIB is lower compared to the LIB, leading to the lower energy density of the SIB (a comparison of the different parameters can be seen in Table 6.2). This problem can be improved by increasing the theoretical capacitance or increasing the cell voltage. Since many of the CMPs used as cathodes have a redox potential of 2~3 V, it is very difficult to design a material with a higher redox potential. On the other hand, when the negative electrode is considered, a material having an oxidation-reduction potential close to 0 V (vs. sodium) is required. In the case of LIBs, anode materials such as graphite (with appropriate redox potential) are well developed, but in the case of SIBs, finding an anode close to the performance of graphite is still a challenge.

6.4 CMPs for Supercapacitors

Supercapacitors are energy storage devices with very high power and low energy density compared to batteries. Each of these properties has advantages and disadvantages. However, researchers are racing to develop devices with power densities like supercapacitors and energy densities like batteries [103, 104]; the CMP is considered the next candidate to achieve this goal.

Thomas et al. reported a supercapacitive energy storage CMP network synthesized by a BH (Buchwald-Hartwig) coupling reaction (Figure 6.5a) [105]. Poly-aminoanthraquinones (PAQs) were found to combine a high surface area (600 m^2 g^{-1}), redox activity of the quinone group, and very good dispersibility in polar solvents. When PAQ was applied as a supercapacitor electrode, it showed a specific capacitance value of 570 F g^{-1}, retained 80~85% capacitance after 600 cycles, and exhibited a columbic efficiency of nearly 100%. To demonstrate the potential of flexible electronics applications, applications to both symmetric and asymmetric devices and flexible devices have been investigated (Figure 6.5b). It was also noted that both the electric double layer and the pseudocapacitance contribute to the specific capacitance of the device.

A nitrogen-rich aza-fused CMP framework was reported for supercapacitor energy storage (Figure 6.5d) [54]. It was observed that the fused aza frameworks are better electrical conductors, have more active sites on the surface due to their higher surface area, and the microporous structure helps the fast diffusion of ions in the charge/discharge process. Combining these advantages, the aza-CMP delivered an excellent capacity of 946 F g^{-1}, significantly outperforming other CMPs at the time (Figure 6.5c). A capacity of 378 F g^{-1} was observed at a high current density of 10 A g^{-1}, which did not decrease over 10,000 cycles.

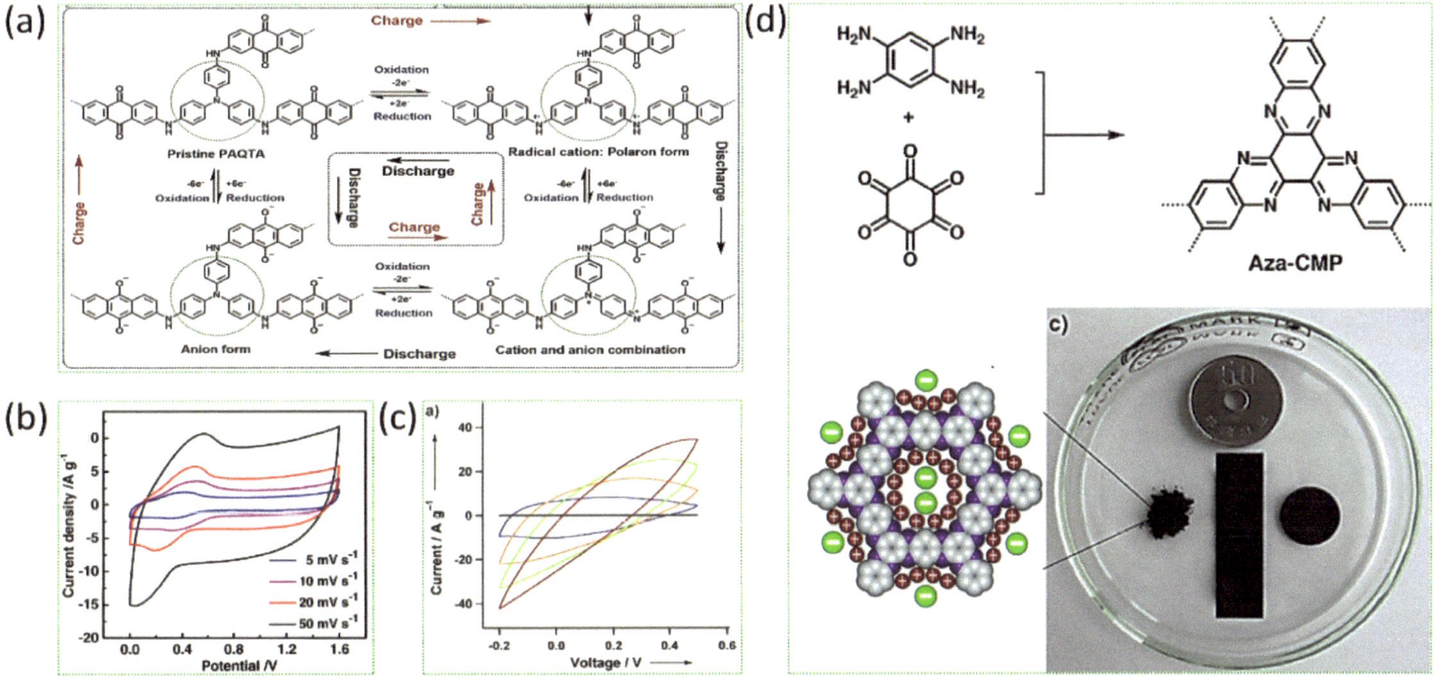

Figure 6.5 (a) Proposed charge discharge mechanism of PAQTA-based supercapacitor and (b) electrochemical performance of PAQTA-activated carbon asymmetric supercapacitor device. Synthesis of (c) CV of a supercapacitor device based on aza-fused CMP and (d) aza-fused CMP. Parts (a) and (b) are reproduced from the literature with permission [105], Copyright © 2018 WILEY-VCH Verlag GmbH & Co. KGaA, Weinheim; (c) and (d) Reproduced with permission from Ref. [54] Copyright © 2011 WILEY-VCH Verlag GmbH & Co. KGaA, Weinheim.

In an interesting study, Chandra et al. discussed the role of hydrogen bonding and the importance of redox-active groups in the energy storage performance of the COFs TpPa-$(OH)_2$ and TpBD-$(OH)_2$ [106]. They concluded that a proton-coupled electron transfer process based on hydroquinone/benzoquinone (H_2Q/Q) was primarily responsible for the high specific capacitance of these materials. When the hydroquinone/benzoquinone moiety was substituted with a phenolic or methoxy group, no redox peak was observed, resulting in a very low specific capacitance. It is also shown that the pseudo capacitance contributes significantly to the total capacitance regardless of the surface area of the material. The hydrogen bonding between HQ and the adjacent amine functional group contributed to the stability and hence the good reversibility of the previously unobserved redox properties.

Two organic monomers, 2, 4, 6-trihydroxyisophthalaldehyde and 1, 3, 5-tris (4-aminophenyl) triazine, were condensed by Schiff base condensation reaction to obtain a new porous polymer TPDA-1 with a high surface area and an extended conjugated network [107]. The authors conclude that ion-conducting channels derived from the microporous network and high surface area are the main causes of this capacitance. High nitrogen doping content and large specific surface area play an important role in pseudocapacitive energy storage processes. In this regard, a conductive covalent triazine framework (CTF) derived from tetracyanoquinodimethane (TCNQ) has been reported to have a nitrogen content of more than 8% and a very high surface area of 3600 $m^2 g^{-1}$ [108].

To improve the conductivity of CMP, graphene hydrogel-based CMP composites have been reported previously [42]. The resulting composite was found to have micro-, meso- and macro-porous networks with a high surface area (219 $m^2 g^{-1}$). When this composite was applied as an electrode of a supercapacitor, a high capacitance of 206 F g^{-1} was obtained. Although the EIS data indicate that the composite has the lowest charge transfer resistance compared to pure CMP, the authors do not provide data on the conductivity of the composite. These papers will provide an overview of how CMP is employed as a capacitive energy storage material in different configurations according to different mechanisms. From the results presented here, nitrogen-doped CMP has the potential to bridge the gap between batteries and supercapacitors (see Table 6.3 for a summary).

Table 6.3 A comparative analysis of different features of CMPs in supercapacitor applications

CMPs	BET ($m^2 g^{-1}$)	Working potential (V)	Specific capacitance ($F g^{-1}$)	Current density ($A g^{-1}$)	Active material loading (% age)	Ref.
PAQTA	331	0–0.6	576	1	80	[105]
Aza-CMP@500	1227	-0.2–0.5	946	0.1	80	[54]
TpPa-$(OH)_2$	369	-0.2–0.5	416	0.5	75	[106]
TPDA-1	545	0.0–1.0	348	0.5	-	[107]
TCNQ=CTA	4000	-0.8–0.1	383	0.2	80	[108]
GH-CMP	219	0.0–1.0	206	0.5	80	[42]

6.5 Conclusion and Future Outlook

Keeping the discussion to this point in mind, we can conclude that conjugated microporous polymers are emerging as strong potential electrode materials for LIBs. It is noteworthy that these materials have already paved the way for a range of different energy storage chemistries like lithium, sulfur, sodium, potassium, magnesium, zinc-based batteries, and supercapacitors [109–111]. Figure 6.6 provides an overview of some of these applications. Also, they can provide us with fascinating structural and dimensional changes by just changing the linker [112]. Solid-state electrolytes are another application where CMPs have already shown their potential [113, 114]. Large surface area coupled with interconnected porous networks and significant conductivity can ensure the fabrication of ultra-fast batteries by enabling efficient charge transfer and ion diffusion processes. Embedding the redox-active organic units into porous frameworks seems a much better strategy for achieving high-performance batteries. However, CMPs are facing several challenges that need to be addressed, including low theoretical capacity and active material utilization, and low conductivity. These factors lead to a reduction in the volumetric energy density of the cell since the high surface area reduces the active material density of the electrode and the porous structure requires more electrolyte to wet the entire electrode. Researchers must therefore carefully balance all these factors when designing CMPs for lithium-ion batteries [109].

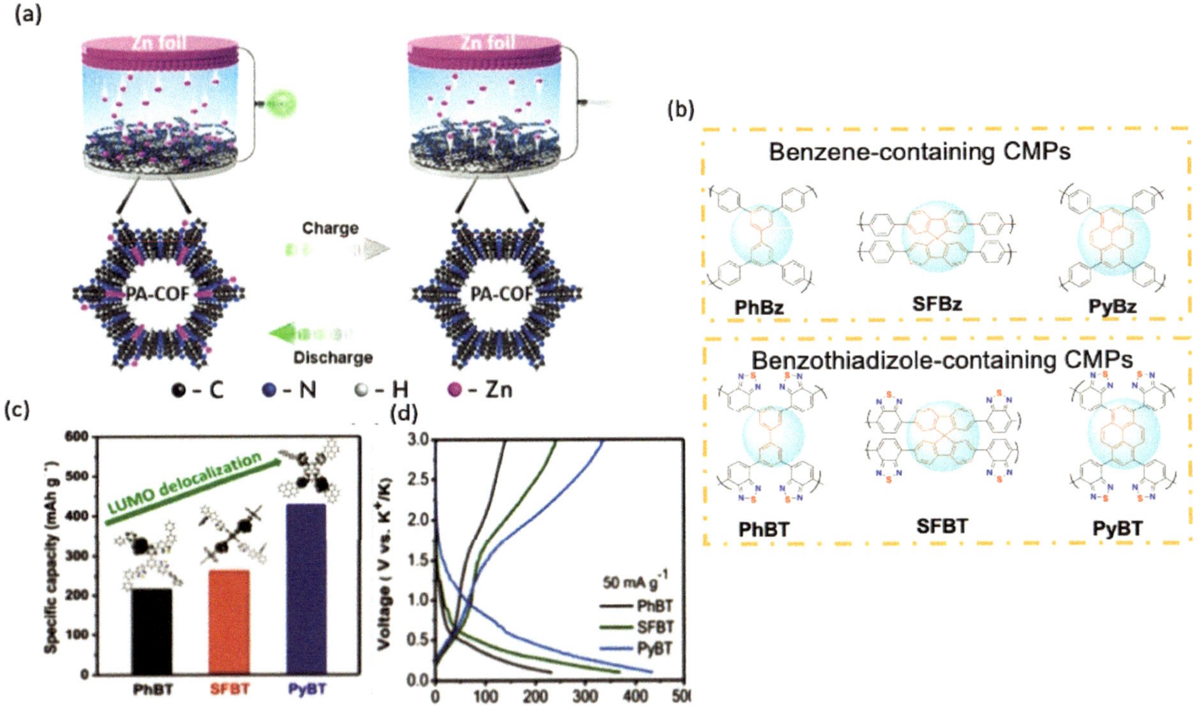

Figure 6.6 (a) Schematic diagram of a zinc-ion ultrafast pseudocapacitive cell, (b) assumed polymer structure of Bz-containing CMP and BT-containing CMP, (c) electronic structure of CMP that plays an important role in battery application, and (d) potassium-ion battery performance of Bz-containing CMP and BT-containing CMP. Part (a) is reproduced from [130] with permission [130], Copyright © 2020, American Chemical Society, (c and d) Reprinted with permission from Ref. [131], Copyright 2019, American Chemical Society.

References

1. E. Pomerantseva, F. Bonaccorso, X. Feng, Y. Cui, Y. Gogotsi, Energy storage: the future enabled by nanomaterials, *Science* 366 (2019), eaan8285.
2. M. Armand, J.M. Tarascon, Building better batteries, *Nature* 451 (2008), 652–657.
3. V. Etacheri, R. Marom, R. Elazari, G. Salitra, D. Aurbach, Challenges in the development of advanced Li-ion batteries: a review, *Energy Environ. Sci.* 4 (2011), 3243–3262.
4. P. Gerlach, A. Balducci, A critical analysis about the underestimated role of the electrolyte in batteries based on organic materials, *ChemElectroChem* 7 (2020), 2364–2375.
5. S. Muench, A. Wild, C. Friebe, B. Haupler, T. Janoschka, U.S. Schubert, Polymer-based organic batteries, *Chem. Rev.* 116 (2016), 9438–9484.
6. K. Turcheniuk, D. Bondarev, V. Singhal, G. Yushin, Ten years left to redesign lithium-ion batteries, *Nature* 559 (2018), 467–470.
7. Q. Cui, Y. Zhong, L. Pan, H. Zhang, Y. Yang, D. Liu, F. Teng, Y. Bando, J. Yao, X. Wang, Recent advances in designing high-capacity anode nanomaterials for li-ion batteries and their atomic-scale storage mechanism studies, *Adv. Sci.* 5 (2018), 1700902.
8. E.A. Olivetti, G. Ceder, G.G. Gaustad, X. Fu, Lithium-ion battery supply chain considerations: analysis of potential bottlenecks in critical metals, *Joule* 1 (2017) 229–243.
9. B. Dunn, H. Kamath, J.-M. Tarascon, Electrical energy storage for the grid: a battery of choices, *Science* 334 (2011), 928–935.
10. L. Oliveira, M. Messagie, S. Rangaraju, J. Sanfelix, M. Hernandez Rivas, J. Van Mierlo, Key issues of lithium-ion batteries – from resource depletion to environmental performance indicators, *J. Clean. Prod.* 108 (2015), 354–362.
11. F. Larsson, P. Andersson, P. Blomqvist, B.E. Mellander, Toxic fluoride gas emissions from lithium-ion battery fires, *Sci. Rep.* 7 (2017), 10018.
12. J.M. Tarascon, M. Armand, Issues and challenges facing rechargeable lithium batteries, *Nature* 414 (2001), 359–367.
13. Y. Zhang, Z. Chen, H. Qiu, W. Yang, Z. Zhao, J. Zhao, G. Cui, Pursuit of reversible Zn electrochemistry: a time-honored challenge towards low-cost and green energy storage, *NPG Asia Mater.* 12 (2020), 4.
14. M. Li, J. Lu, Z. Chen, K. Amine, 30 Years of lithium-ion batteries, *Adv. Mater.* 30 (2018), 1800561.

15. N. Nitta, F. Wu, J.T. Lee, G. Yushin, Li-ion battery materials: present and future, *Mater. Today* 18 (2015), 252–264.

16. Q. Pang, X. Liang, C.Y. Kwok, L.F. Nazar, Advances in lithium–sulfur batteries based on multifunctional cathodes and electrolytes, *Nat. Energy* 1 (2016), 16132.

17. S.-W. Kim, D.-H. Seo, X. Ma, G. Ceder, K. Kang, Electrode materials for rechargeable sodium-ion batteries: potential alternatives to current lithium-ion batteries, *Adv. Energy Mater.* 2 (2012), 710–721.

18. T. Yokoji, H. Matsubara, M. Satoh, Rechargeable organic lithium-ion batteries using electron-deficient benzoquinones as positive-electrode materials with high discharge voltages, *J. Mater. Chem. A* 2 (2014), 19347–19354.

19. T.B. Schon, B.T. McAllister, P.F. Li, D.S. Seferos, The rise of organic electrode materials for energy storage, *Chem. Soc. Rev.* 45 (2016), 6345–6404.

20. B. H¨ aupler, A. Wild, U.S. Schubert, Carbonyls: powerful organic materials for secondary batteries, *Adv. Energy Mater.* 5 (2015), 1402034.

21. Y. Wu, R. Zeng, J. Nan, D. Shu, Y. Qiu, S.-L. Chou, Quinone electrode materials for rechargeable lithium/sodium ion batteries, *Adv. Energy Mater.* 7 (2017), 1700278.

22. H. Oubaha, J.F. Gohy, S. Melinte, Carbonyl-based π-conjugated materials: from synthesis to applications in lithium-ion batteries, *ChemPlusChem* 84 (2019), 1179–1214.

23. Y. Liang, Z. Tao, J. Chen, Organic electrode materials for rechargeable lithium batteries, *Adv. Energy Mater.* 2 (2012), 742–769.

24. K. Amin, L. Mao, Z. Wei, Recent Progress in polymeric carbonyl-based electrode materials for lithium and sodium ion batteries, *Macromol. Rapid Commun.* 40 (2019), 1800565.

25. H. Wu, Q. Meng, Q. Yang, M. Zhang, K. Lu, Z. Wei, Large-area polyimide/SWCNT nanocable cathode for flexible lithium-ion batteries, *Adv. Mater.* 27 (2015), 6504–6510.

26. K. Amin, Q. Meng, A. Ahmad, M. Cheng, M. Zhang, L. Mao, K. Lu, Z. Wei, A carbonyl compound-based flexible cathode with superior rate performance and cyclic stability for flexible lithium-ion batteries, *Adv. Mater.* 30 (2018), 1703868.

27. H.P. Wu, Q. Yang, Q.H. Meng, A. Ahmad, M. Zhang, L.Y. Zhu, Y.G. Liu, Z.X. Wei, A polyimide derivative containing different carbonyl groups for flexible lithium-ion batteries, *J. Mater. Chem. A* 4 (2016), 2115–2121.

28. Y. Meng, H. Wu, Y. Zhang, Z. Wei, A flexible electrode based on a three-dimensional graphene network-supported polyimide for lithium-ion batteries, *J. Mater. Chem. A* 2 (2014), 10842–10846.

29. Y. Zhang, S.N. Riduan, J. Wang, Redox active metal- and covalent organic frameworks for energy storage: balancing porosity and electrical conductivity, *Chem. (Easton)* 23 (2017), 16419–16431.

30. Y. Xu, S. Jin, H. Xu, A. Nagai, D. Jiang, Conjugated microporous polymers: design, synthesis and application, *Chem. Soc. Rev.* 42 (2013), 8012–8031.

31. J.M. Lee, A.I. Cooper, Advances in conjugated microporous polymers, *Chem. Rev.* 120 (2020), 2171–2214.

32. N. Chaoui, M. Trunk, R. Dawson, J. Schmidt, A. Thomas, Trends and challenges for microporous polymers, *Chem. Soc. Rev.* 46 (2017), 3302–3321.

33. A.I. Cooper, Conjugated microporous polymers, *Adv. Mater.* 21 (2009), 1291–1295.

34. J. Chen, T. Qiu, W. Yan, C.F.J. Faul, Exploiting Hansen solubility parameters to tune porosity and function in conjugated microporous polymers, *J. Mater. Chem. A* 8 (2020), 22657–22665.

35. J. Chen, W. Yan, E.J. Townsend, J. Feng, L. Pan, V. Del Angel Hernandez, C.F. J. Faul, Tunable surface area, porosity, and function in conjugated microporous polymers, *Angew. Chem. Int. Ed. Engl.* 58 (2019), 11715–11719.

36. B. Atsbeha Kahsay, F.M. Wang, A.G. Hailu, C.H. Su, Maleamic acid as an organic anode material in lithium-ion batteries, *Polymers* 12 (2020), 1109.

37. Q. Zhang, S. Yu, Q. Wang, Q. Xiao, Y. Yue, S. Ren, Fluorene-based conjugated microporous polymers: preparation and chemical sensing application, *Macromol. Rapid Commun.* 38 (2017), 1700445.

38. S. Luo, Z. Zeng, G. Zeng, Z. Liu, R. Xiao, P. Xu, H. Wang, D. Huang, Y. Liu, B. Shao, Q. Liang, D. Wang, Q. He, L. Qin, Y. Fu, Recent advances in conjugated microporous polymers for photocatalysis: designs, applications, and prospects, *J. Mater. Chem. A* 8 (2020), 6434–6470.

39. J. Tan, W.-J. Chen, J. Guo, Conjugated microporous polymers with distinctive π-electronic properties exhibiting enhanced optical applications, *Chin. Chem. Lett.* 27 (2016), 1405–1411.

40. N. Chaoui, M. Trunk, R. Dawson, J. Schmidt, A. Thomas, Trends and challenges for microporous polymers, *Chem. Soc. Rev.* 46 (2017), 3302–3321.

41. Y.B. Zhou, Z.P. Zhan, Conjugated microporous polymers for heterogeneous catalysis, *Chem. Asian J.* 13 (2018), 9–19.
42. M. Zhang, T. Zhao, J. Dou, Z. Xu, W. Zhang, X. Chen, X. Wang, B. Zhou, Bottom-up construction of conjugated microporous polyporphyrin-coated graphene hydrogel composites with hierarchical pores for high-performance capacitors, *ChemElectroChem* 6 (2019), 5946–5950.
43. M.E. Bhosale, R. Illathvalappil, S. Kurungot, K. Krishnamoorthy, Conjugated porous polymers as precursors for electrocatalysts and storage electrode materials, *Chem. Commun.* 52 (2016), 316–318.
44. L. Pan, Z. Liu, M. Tian, B.C. Schroeder, A.E. Aliev, C.F.J. Faul, Luminescent and swellable conjugated microporous polymers for detecting nitroaromatic explosives and removing harmful organic vapors, *ACS Appl. Mater. Interfaces* 11 (2019), 48352–48362.
45. Y. You, A. Manthiram, Progress in high-voltage cathode materials for rechargeable sodium-ion batteries, *Adv. Energy Mater.* 8 (2018), 1701785.
46. C.-S. Li, Y. Sun, F. Gebert, S.-L. Chou, Current progress on rechargeable magnesium-air battery, *Adv. Energy Mater.* 7 (2017), 1700869.
47. J. Fu, Z.P. Cano, M.G. Park, A. Yu, M. Fowler, Z. Chen, Electrically rechargeable zinc-air batteries: progress, challenges, and perspectives, *Adv. Mater.* 29 (2017), 1604685.
48. J. Qin, H.M. Kheimeh Sari, C. He, X. Li, A hybrid energy storage mechanism of carbonous anodes harvesting superior rate capability and long cycle life for sodium/potassium storage, *J. Mater. Chem. A* 7 (2019), 3673–3681.
49. R. Liu, K.T. Tan, Y. Gong, Y. Chen, Z. Li, S. Xie, T. He, Z. Lu, H. Yang, D. Jiang, Covalent organic frameworks: an ideal platform for designing ordered materials and advanced applications, *Chem. Soc. Rev.* 50 (2020), 120–242.
50. X. Liu, C.F. Liu, W.Y. Lai, W. Huang, Porous organic polymers as promising electrode materials for energy storage devices, *Adv. Mater. Technol.* 5 (2020).
51. T. Sun, J. Xie, W. Guo, D.S. Li, Q.C. Zhang, Covalent-organic frameworks: advanced organic electrode materials for rechargeable batteries, *Adv. Energy Mater.* 10 (2020), 1904199.
52. P. Nov′ ak, K. Müller, K. Santhanam, O. Haas, Electrochemically active polymers for rechargeable batteries, *Chem. Rev.* 97 (1997), 207–282.
53. X. Han, C. Chang, L. Yuan, T. Sun, J. Sun, Aromatic carbonyl derivative polymers as high-performance Li-ion storage materials, *Adv. Mater.* 19 (2007), 1616–1621.

54. Y. Kou, Y. Xu, Z. Guo, D. Jiang, Supercapacitive energy storage and electric power supply using an aza-fused pi-conjugated microporous framework, *Angew. Chem. Int. Ed. Engl.* 50 (2011), 8753–8757.

55. J.L. Segura, M.J. Mancheno, F. Zamora, Covalent organic frameworks based on Schiff-base chemistry: synthesis, properties and potential applications, *Chem. Soc. Rev.* 45 (2016), 5635–5671.

56. Y. Liao, Z. Cheng, W. Zuo, A. Thomas, C.F.J. Faul, Nitrogen-rich conjugated microporous polymers: facile synthesis, efficient gas storage, and heterogeneous catalysis, *ACS Appl. Mater. Interfaces* 9 (2017), 38390–38400.

57. Y. Liao, J. Weber, C.F. Faul, Conjugated microporous polytriphenylamine networks, *Chem. Commun. (Camb.)* 50 (2014), 8002–8005.

58. S.J. Lyle, P.J. Waller, O.M. Yaghi, Covalent organic frameworks: organic chemistry extended into two and three dimensions, *Trends Chem.* 1 (2019), 172–184.

59. C. Zhang, X. Yang, W. Ren, Y. Wang, F. Su, J.-X. Jiang, Microporous organic polymer-based lithium-ion batteries with improved rate performance and energy density, *J. Power Sources* 317 (2016), 49–56.

60. M. Wu, Y. Zhao, B. Sun, Z. Sun, C. Li, Y. Han, L. Xu, Z. Ge, Y. Ren, M. Zhang, Q. Zhang, Y. Lu, W. Wang, Y. Ma, Y. Chen, A 2D covalent organic framework as a high-performance cathode material for lithium-ion batteries, *Nano Energy* 70 (2020), 104498.

61. C. Peng, G.-H. Ning, J. Su, G. Zhong, W. Tang, B. Tian, C. Su, D. Yu, L. Zu, J. Yang, M.-F. Ng, Y.-S. Hu, Y. Yang, M. Armand, K.P. Loh, Reversible multi-electron redox chemistry of π-conjugated N-containing heteroaromatic molecule-based organic cathodes, *Nat. Energy* 2 (2017), 17074.

62. K. Nakashima, T. Shimizu, Y. Kamakura, A. Hinokimoto, Y. Kitagawa, H. Yoshikawa, D. Tanaka, A new design strategy for redox-active molecular assemblies with crystalline porous structures for lithium-ion batteries, *Chem. Sci.* 11 (2020), 37–43.

63. Z. Chen, W. Li, Y. Dai, N. Xu, C. Su, J. Liu, C. Zhang, Conjugated microporous polymer based on star-shaped triphenylamine-benzene structure with improved electrochemical performances as the organic cathode material of Li-ion battery, *Electrochim. Acta* 286 (2018), 187–194.

64. Z. Luo, L. Liu, J. Ning, K. Lei, Y. Lu, F. Li, J. Chen, A microporous covalent-organic framework with abundant accessible carbonyl groups for lithium-ion batteries, *Angew. Chem. Int. Ed. Engl.* 57 (2018), 9443–9446.

65. C.R. DeBlase, K. Hernandez-Burgos, J.M. Rotter, D.J. Fortman, S. Abreu Ddos, R. A. Timm, I.C. Diogenes, L.T. Kubota, H.D. Abruna, W.R. Dichtel, Cation- dependent stabilization of electrogenerated naphthalene diimide dianions in porous polymer thin films and their application to electrical energy storage, *Angew. Chem. Int. Ed. Engl.* 54 (2015), 13225–13229.

66. S. Wang, Q. Wang, P. Shao, Y. Han, X. Gao, L. Ma, S. Yuan, X. Ma, J. Zhou, X. Feng, B. Wang, Exfoliation of covalent organic frameworks into few-layer redox-active nanosheets as cathode materials for lithium-ion batteries, *J. Am. Chem. Soc.* 139 (2017), 4258–4261.

67. C.R. Mulzer, L. Shen, R.P. Bisbey, J.R. McKone, N. Zhang, H.D. Abruna, W. R. Dichtel, Superior charge storage and power density of a conducting polymer-modified covalent organic framework, *ACS Cent. Sci.* 2 (2016), 667–673.

68. Y. Wu, D. Yan, Z. Zhang, M.M. Matsushita, K. Awaga, Electron highways into nanochannels of covalent organic frameworks for high electrical conductivity and energy storage, *ACS Appl. Mater. Interfaces* 11 (2019), 7661–7665.

69. E. Vitaku, C.N. Gannett, K.L. Carpenter, L. Shen, H.D. Abruna, W.R. Dichtel, Phenazine-based covalent organic framework cathode materials with high energy and power densities, *J. Am. Chem. Soc.* 142 (2019), 16–20.

70. K. Amin, J.Q. Zhang, H.Y. Zhou, R.C. Lu, Z. Miao, N. Ashraf, Y.L. Cheng, L.J. Mao, C.F.J. Faut, Z.X. Wei, Surface controlled pseudo-capacitive reactions enabling ultra-fast charging and long-life organic lithium-ion batteries, *Sustain. Energy Fuels* 4 (2020), 4179–4185.

71. S. Xu, G. Wang, B.P. Biswal, M. Addicoat, S. Paasch, W. Sheng, X. Zhuang, E. Brunner, T. Heine, R. Berger, X. Feng, A. Nitrogen-Rich, 2D sp(2)-carbon-linked conjugated polymer framework as a high-performance cathode for lithium-ion batteries, *Angew. Chem. Int. Ed. Engl.* 58 (2019), 849–853.

72. X. Li, H. Wang, H. Chen, Q. Zheng, Q. Zhang, H. Mao, Y. Liu, S. Cai, B. Sun, C. Dun, M.P. Gordon, H. Zheng, J.A. Reimer, J.J. Urban, J. Ciston, T. Tan, E. M. Chan, J. Zhang, Y. Liu, Dynamic covalent synthesis of crystalline porous graphitic frameworks, *Chem* 6 (2020), 933–944.

73. T. Banerjee, B.V. Lotsch, Holey heterographenes made to order: "green" synthesis of porous graphitic frameworks, *Chem* 6 (2020), 812–814.

74. J. Wang, J.C.Z. En, S.N. Riduan, Y. Zhang, Nitrogen-linked hexaazatrinaphthylene polymer as cathode material in lithium-ion battery, *Chemistry* 26 (2020), 2581–2585.

75. S. Zhang, W. Huang, P. Hu, C. Huang, C. Shang, C. Zhang, R. Yang, G. Cui, Conjugated microporous polymers with excellent electrochemical performance for lithium and sodium storage, *J. Mater. Chem. A* 3 (2015), 1896–1901.

76. T. Yang, C. Zhang, W. Ma, X. Gao, C. Yan, F. Wang, J.-X. Jiang, Thiophene-rich conjugated microporous polymers as anode materials for high performance lithium- and sodium-ion batteries, *Solid State Ion.* 347 (2020), 115247.

77. Y. Zhu, X. Chen, Y. Cao, W. Peng, Y. Li, G. Zhang, F. Zhang, X. Fan, Reversible intercalation and exfoliation of layered covalent triazine frameworks for enhanced lithium-ion storage, *Chem. Commun.* 55 (2019), 1434–1437.

78. Z. Lei, Q. Yang, Y. Xu, S. Guo, W. Sun, H. Liu, L.P. Lv, Y. Zhang, Y. Wang, Boosting lithium storage in covalent organic framework via activation of 14-electron redox chemistry, *Nat. Commun.* 9 (2018), 576.

79. V. Palomares, P. Serras, I. Villaluenga, K.B. Hueso, J. Carretero-Gonz´ alez, T. Rojo, Na-ion batteries, recent advances and present challenges to become low cost energy storage systems, *Energy Environ. Sci.* 5 (2012), 5884–5901.

80. S.W. Kim, D.H. Seo, X. Ma, G. Ceder, K. Kang, Electrode materials for rechargeable sodium-ion batteries: potential alternatives to current lithium-ion batteries, *Adv. Energy Mater.* 2 (2012), 710–721.

81. R. Shi, L. Liu, Y. Lu, C. Wang, Y. Li, L. Li, Z. Yan, J. Chen, Nitrogen-rich covalent organic frameworks with multiple carbonyls for high-performance sodium batteries, *Nat. Commun.* 11 (2020), 178.

82. J.X. Jiang, C. Wang, A. Laybourn, T. Hasell, R. Clowes, Y.Z. Khimyak, J. Xiao, S. J. Higgins, D.J. Adams, A.I. Cooper, Metal–organic conjugated microporous polymers, *Angew. Chem. Int. Ed.* 50 (2011), 1072–1075.

83. C. Zhang, Y. Qiao, P. Xiong, W. Ma, P. Bai, X. Wang, Q. Li, J. Zhao, Y. Xu, Y. Chen, Conjugated microporous polymers with tunable electronic structure for high-performance potassium-ion batteries, *ACS Nano* 13 (2019), 745–754.

84. Y. Park, D.S. Shin, S.H. Woo, N.S. Choi, K.H. Shin, S.M. Oh, K.T. Lee, S.Y. Hong, Sodium terephthalate as an organic anode material for sodium ion batteries, *Adv. Mater.* 24 (2012), 3562–3567.

85. L. Zhao, J. Zhao, Y.S. Hu, H. Li, Z. Zhou, M. Armand, L. Chen, Disodium terephthalate ($Na_2C_8H_4O_4$) as high performance anode material for low-cost room-temperature sodium-ion battery, *Adv. Energy Mater.* 2 (2012), 962–965,

86. P.K. Nayak, L. Yang, W. Brehm, P. Adelhelm, From lithium-ion to sodium-ion batteries: advantages, challenges, and surprises, *Angew. Chem. Int. Ed. Engl.* 57 (2018), 102–120.

87. K. Sakaushi, E. Hosono, G. Nickerl, T. Gemming, H. Zhou, S. Kaskel, J. Eckert, Aromatic porous-honeycomb electrodes for a sodium-organic energy storage device, *Nat. Commun.* 4 (2013), 1–7.

88. Z. Wu, J. Xie, Z.J. Xu, S. Zhang, Q. Zhang, Recent progress in metal–organic polymers as promising electrodes for lithium/sodium rechargeable batteries, *J. Mater. Chem. A* 7 (2019), 4259–4290.

89. T.B. Schon, B.T. McAllister, P.-F. Li, D.S. Seferos, The rise of organic electrode materials for energy storage, *Chem. Soc. Rev.* 45 (2016), 6345–6404.

90. R.B. dos Santos, R. Rivelino, F. d B. Mota, G.K. Gueorguiev, Effects of N doping on the electronic properties of a small carbon atomic chain with distinct sp2 terminations: a first-principles study, *Phys. Rev. B* 84 (2011), 075417.

91. M. Shakourian-Fard, H.R. Ghenaatian, G. Kamath, S.M. Taimoory, Unraveling the effect of nitrogen doping on graphene nanoflakes and the adsorption properties of ionic liquids: a DFT study, *J. Mol. Liq.* 312 (2020), 113400.

92. B. Quan, A. Jin, S.H. Yu, S.M. Kang, J. Jeong, H.D. Abruna, L. Jin, Y. Piao, Y. E. Sung, Solvothermal-derived s-doped graphene as an anode material for sodium-ion batteries, *Adv. Sci.* 5 (2018), 1700880.

93. X. Deng, K. Xie, L. Li, W. Zhou, J. Sunarso, Z. Shao, Scalable synthesis of self-standing sulfur-doped flexible graphene films as recyclable anode materials for low-cost sodium-ion batteries, *Carbon* 107 (2016), 67–73.

94. K.S. Weeraratne, A.A. Alzharani, H.M. El-Kaderi, Redox-active porous organic polymers as novel electrode materials for green rechargeable sodium-ion batteries, *ACS Appl. Mater. Interfaces* 11 (2019), 23520–23526.

95. H. Li, M. Tang, Y. Wu, Y. Chen, S. Zhu, B. Wang, C. Jiang, E. Wang, C. Wang, Large pi-conjugated porous frameworks as cathodes for sodium-ion batteries, *J. Phys. Chem. Lett.* 9 (2018), 3205–3211.

96. J.-K. Sun, Q. Xu, Functional materials derived from open framework templates/precursors: synthesis and applications, *Energy Environ. Sci.* 7 (2014), 2071–2100.

97. K. Sasidharachari, B.-K. Na, S.-G. Woo, S. Yoon, K.Y. Cho, Facile conductive surface modification of Si nanoparticle with nitrogen-doped carbon

layers for lithium-ion batteries, *J. Solid State Electrochem.* 20 (2016), 2873–2878.

98. S. Wang, L. Shang, L. Li, Y. Yu, C. Chi, K. Wang, J. Zhang, R. Shi, H. Shen, G. I. Waterhouse, Metal–organic-framework-derived mesoporous carbon nanospheres containing porphyrin-like metal centers for conformal phototherapy, *Adv. Mater.* 28 (2016), 8379–8387.

99. X. Zhang, G. Zhu, M. Wang, J. Li, T. Lu, L. Pan, Covalent-organic-frameworks derived N-doped porous carbon materials as anode for superior long-life cycling lithium and sodium ion batteries, *Carbon* 116 (2017), 686–694.

100. S. Gu, S. Wu, L. Cao, M. Li, N. Qin, J. Zhu, Z. Wang, Y. Li, Z. Li, J. Chen, Tunable redox chemistry and stability of radical intermediates in 2D covalent organic frameworks for high performance sodium ion batteries, *J. Am. Chem. Soc.* 141 (2019), 9623–9628.

101. M.-S. Kim, W.-J. Lee, S.-M. Paek, J.K. Park, Covalent organic nanosheets as effective sodium-ion storage materials, *ACS Appl. Mater. Interfaces* 10 (2018), 32102–32111.

102. Y. Li, B. Ni, X. Li, X. Wang, D. Zhang, Q. Zhao, J. Li, T. Lu, W. Mai, L. Pan, High-performance na-ion storage of s-doped porous carbon derived from conjugated microporous polymers, *Nanomicro Lett.* 11 (2019), 60.

103. R. Thangavel, K. Kaliyappan, D.-U. Kim, X. Sun, Y.-S. Lee, All-organic sodium hybrid capacitor: a new, high-energy, high-power energy storage system bridging batteries and capacitors, *Chem. Mater.* 29 (2017), 7122–7130.

104. V. Augustyn, P. Simon, B. Dunn, Pseudocapacitive oxide materials for high-rate electrochemical energy storage, *Energy Environ. Sci.* 7 (2014), 1597.

105. Y. Liao, H. Wang, M. Zhu, A. Thomas, Efficient supercapacitor energy storage using conjugated microporous polymer networks synthesized from Buchwald- Hartwig coupling, *Adv. Mater.* 30 (2018), 1705710.

106. S. Chandra, D. Roy Chowdhury, M. Addicoat, T. Heine, A. Paul, R. Banerjee, Molecular level control of the capacitance of two-dimensional covalent organic frameworks: role of hydrogen bonding in energy storage materials, *Chem. Mater.* 29 (2017), 2074–2080.

107. P. Bhanja, S.K. Das, K. Bhunia, D. Pradhan, T. Hayashi, Y. Hijikata, S. Irle, A. Bhaumik, A new porous polymer for highly efficient capacitive energy storage, *ACS Sustain. Chem. Eng.* 6 (2017), 202–209.

108. Y. Li, S. Zheng, X. Liu, P. Li, L. Sun, R. Yang, S. Wang, Z.S. Wu, X. Bao, W.Q. Deng, Conductive microporous covalent triazine-based framework for high-performance electrochemical capacitive energy storage, *Angew. Chem. Int. Ed. Engl.* 57 (2018), 7992–7996.

109. W. Wang, V.S. Kale, Z. Cao, S. Kandambeth, W. Zhang, J. Ming, P.T. Parvatkar, E. Abou-Hamad, O. Shekhah, L. Cavallo, M. Eddaoudi, H.N. Alshareef, Phenanthroline covalent organic framework electrodes for high-performance zinc-ion supercapattery, *ACS Energy Lett.* 5 (2020), 2256–2264.

110. C. Zhang, Y. Qiao, P. Xiong, W. Ma, P. Bai, X. Wang, Q. Li, J. Zhao, Y. Xu, Y. Chen, J.H. Zeng, F. Wang, Y. Xu, J.X. Jiang, Conjugated microporous polymers with tunable electronic structure for high-performance potassium-ion batteries, *ACS Nano* 13 (2019), 745–754.

111. Z. Li, H.-Y. Zhou, F.-L. Zhao, T.-X. Wang, X. Ding, B.-H. Han, W. Feng, Three-dimensional covalent organic frameworks as host materials for lithium-sulfur batteries, *Chin. J. Polym. Sci.* 38 (2020), 550–557.

112. Z. Li, X. Ding, Y. Feng, W. Feng, B.-H. Han, Structural and dimensional transformations between covalent organic frameworks via linker exchange, *Macromolecules* 52 (2019), 1257–1265.

113. Z. Li, Z.-W. Liu, Z.-J. Mu, C. Cao, Z. Li, T.-X. Wang, Y. Li, X. Ding, B.-H. Han, W. Feng, Cationic covalent organic framework based all-solid-state electrolytes, *Mater. Chem. Front.* 4 (2020), 1164–1173.

114. Z. Li, Z.W. Liu, Z. Li, T.X. Wang, F. Zhao, X. Ding, W. Feng, B.H. Han, Defective 2D covalent organic frameworks for postfunctionalization, *Adv. Funct. Mater.* 30 (2020).

Chapter 7

CMP's Semiconductor Applications

There has been considerable focus on the development of charge transfer (CT) complex formation to modify the bandgaps of organic materials. In particular, CT complexes alternate layers of aromatic molecules with donor (D) and acceptor (A) properties to provide inherent electrical conductivity. In particular, the synthetic porous frameworks as attractive D-A components have been extensively studied in recent years compared to existing D-A materials. Therefore, in this chapter, the synthetic porous frameworks into conjugated microporous polymers (CMPs), covalent organic frameworks (COFs), and metal-organic frameworks (MOFs) are classified, and high-quality materials for CT in semiconductors are compared. Furthermore, several theoretical investigations, efficiency improvement techniques, and a discussion of the electrical conductivity of the porous frameworks are also highlighted. Finally, a perspective of synthetic porous framework studies on CT performance is provided along with some comparisons.

7.1 Porous Materials

Since the discovery of synthetic porous materials, researchers from all over the world have made significant contributions to their chemistry, leading to a considerable growth in publications. Synthetic

Conjugated Microporous Polymers
Atsushi Nagai

ISBN 978-981-5129-66-3 (Hardcover), 978-1-003-63652-6 (eBook)
www.jennystanford.com

organic frameworks have attracted attention as a promising class of porous materials because of their well-organized structures and uniform porosity. Therefore, they have become fundamental building blocks in the development of innovative organic porous materials with significant charge transfer (CT) applications [1]. It is necessary to evaluate the state of recently developed synthetic porous materials after 10 years of development, which can be employed for a variety of CT applications, particularly in semiconductors. These porous organic frameworks can be classified based on various chemical conjugations, including π-conjugated microporous polymers (CMPs) [2], covalently-bonded organic frameworks (COFs) [3], and metal coordination-bonded organic frameworks (MOFs) [4]. These synthetic porous materials with numerous combinations of organic building blocks possess unique properties, advantages, and disadvantages. In addition, comprehensive structural analysis and well-specified organic framework structures provide a solid foundation for additional applications [1d]. CMPs are a class of porous materials comprising extended networks of π-conjugated (alternating single and multiple) bonds. These amorphous materials have a high degree of π-electron delocalization along their backbone [1b, 5]. COFs are a type of porous crystalline material composed of organic molecules connected by strong covalent bonds. These materials have a well-defined structure with molecular-sized pores, offering a high surface area and tunable chemical functionalities [3b, 6]. MOFs are a class of porous materials consisting of metal ions or clusters coordinated to organic ligands, forming a three-dimensional network with well-defined pores. MOFs exhibit high surface areas and diverse structures due to the combination of inorganic metal nodes and organic linkers. Over 20,000 different MOFs have been explored and analyzed in the last 10 years due to the ability to customize constituents, geometry, size, and functionality [7].

Recently, there has been extensive research interest in the utilization of organic bulk heterojunctions that integrate electron donor (D) and acceptor (A) phases, particularly in semiconductor applications [2]. Furthermore, the development of tunable porous materials creates model systems characterized by periodic and interpenetrating networks of electron-D and A phases. This degree of control enables precise manipulation of both structural and electronic properties, maximizing the potential of such materials

[8]. A wide range of CT complexes based on CMP structures from synthetic building blocks and network-forming reactions have been developed for applications such as photocatalysis, energy storage, sensing, and detection, electrocatalysis, molecular electronics, and electronic devices (Figure 7.1) [9]. Additionally, over the past 10 years, significant attention has been paid to the outstanding properties of COFs, such as their exceptional porosity, great stability, enduring porosity, low density, conjugated framework, and customizable functionality. These properties demonstrate the potential of COFs in facilitating CT processes across various fields, including electronics, energy storage, photovoltaics, catalysis, electrocatalysis, and sensing (Figure 7.1) [1a, 10]. The precise manipulation of COF structures and properties provides an opportunity to tailor CT behavior, paving the way for the development of advanced functional materials with unprecedented capabilities. Furthermore, MOFs serve as attractive scaffolding due to their tunable structure pores, high surface area, hierarchical porous structure, high chemical, and good thermal stability. Therefore, CT complexes formed through MOFs with the mentioned features can be used in applications such as electrocatalysis [11], energy storage [12], photocatalysts, chemical-sensing elements, photoelectrochemical (PEC) energy conversion, electrical conductivity, and nanometer-scale objects (Figure 7.1) [13].

Therefore, it is time to investigate a comprehensive review and analysis of synthetic porous frameworks as a platform for semiconductors of CT complexes. In recent years, synthetic porous frameworks have been extensively studied as desirable D-A components compared to current D-A materials. Hence, this review categorizes synthetic porous frameworks as CMPs, COFs, and MOFs to compare high-quality materials for CT in semiconductors. The purpose of this study is to provide the latest trends in CT based on CMPs, COFs, and MOFs, with a specific focus on semiconductor applications in gas storage, absorption, photocatalysis, energy systems, supercapacitors, fuel cells, and photovoltaic cells. Additionally, the review presents an updated overview of research development in these areas. First, the study focuses on the bibliography and history of CMPs, COFs, and MOFs as semiconductors. Next, a comprehensive discussion on CT provides an overview and concepts of theoretical strategies employed in CT studies. Then, an introduction to CMPs, COFs, and MOFs and a comparative analysis of CT performance for porous framework applications are presented. After that, the synthesis and functionalization of CT complexes using CMP, COF, and MOF are outlined. Additionally, the review explores several theoretical studies highlighting the exceptional properties of these materials in the field of CT. The final section examines potential advances and major challenges in the development of porous frameworks for CT applications.

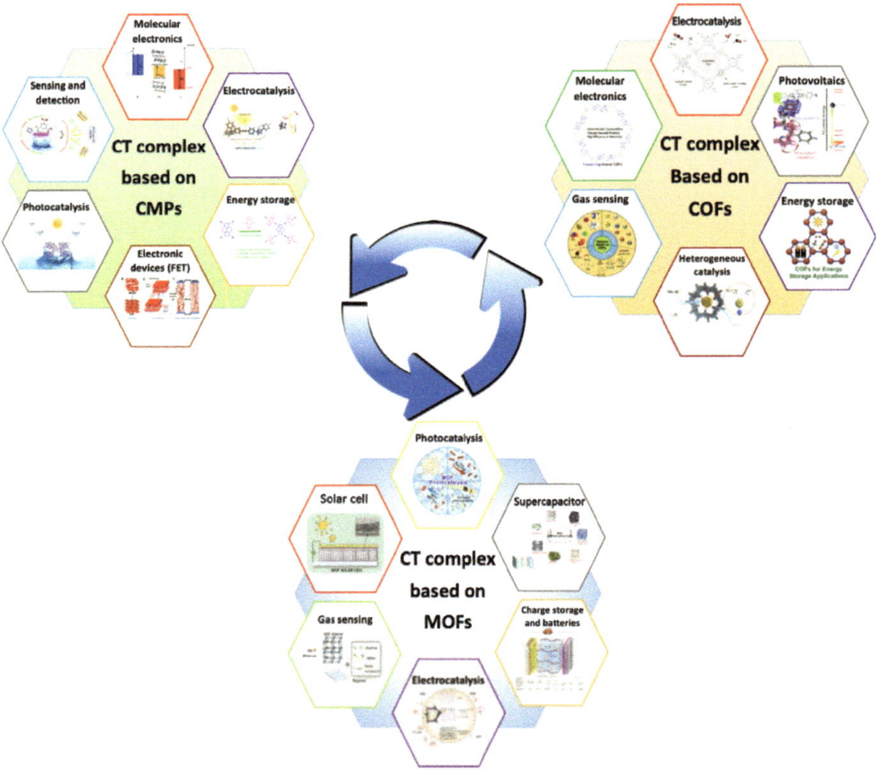

Figure 7.1 Applications of CT complex based on CMPs, COFs, and MOFs.

7.2 Charge Transfer of Concept

Emissive CT refers to the process in which an excited state is generated due to the transfer of an electron from one molecular entity (D) to another (A). In this process, the energy difference between the ground state and the CT state is released as light, resulting in emission. The subsequent step is CT to excited state electron transfer conversion, where the CT state transforms into an excited state. This transformation can lead to either fluorescence or phosphorescence, depending on the spin characteristics of the excited state. Efficient CT to ET conversion is vital for maximizing the emission efficiency and overall performance of optoelectronic devices. In other words, the orientation of transition dipoles plays a significant role in CT processes. A transition dipole moment represents the change in the distribution of electronic charge in a molecule during a transition, such as the shift from the ground state to an excited state. In CT systems, the relative orientation of transition dipoles between the D and A molecules greatly influences the efficiency of CT and subsequent emission. When the transition dipoles of the D and A are parallel, the CT process is typically more efficient. This parallel alignment allows for better electronic coupling between the D and A states, facilitating faster CT and leading to enhanced emission efficiency. In contrast, a perpendicular orientation hampers efficient CT. When the transition dipoles are perpendicular, the electronic coupling between the D and A states is weaker, leading to slower CT rates and decreased emission efficiency. Understanding and controlling the orientation of transition dipoles in CT systems are essential for optimizing the design of organic electronic devices. Researchers often employ molecular engineering techniques to align the transition dipoles, ensuring a parallel orientation and thereby maximizing the efficiency of CT and subsequent emission processes. By manipulating the molecular structure and intermolecular interactions, scientists can fine-tune the orientation of transition dipoles, leading to improved performance in a wide range of optoelectronic applications [14].

In addition, Jablonski introduced the initial approach utilized in organic photochemistry. The Jablonski diagram depicts the energy levels of molecules. When a photon with sufficient energy is absorbed, as illustrated in Figure 7.2, an electron can be promoted to an excited state. In the context of CT, this excitation can lead to an electron being transferred from the D molecule to the A molecule. Light absorption occurs within an incredibly brief femtosecond timeframe, resulting in the excitation of the particle from the ground state (S_0) to an excited state (S_1, S_2, ...). In some cases, the excited electron moves from the D molecule (S_1) to the A molecule (S_2), creating charged species (anions and cations). Each state has specific vibrational levels that become populated upon excitation with varying probabilities. The combination of these levels constitutes an absorption spectrum. Following photon absorption, the most likely process is termed internal conversion or vibrational relaxation. This process, occurring over a longer duration (in the picosecond timeframe), is accompanied by a structural relaxation of the excited molecule. The molecule can remain in this excited state for nanoseconds or even longer before returning to the ground state. During this transition, it emits a photon, a phenomenon known as fluorescence. Following excitation, various events may unfold, such as non-radiative relaxation due to collisions with other particles or intersystem crossing to the lowest excited triplet state (T_1). Relaxation from the triplet state to the ground state, accompanied by photon emission, is known as phosphorescence. Alternatively, a transition back to the S_1 state is possible, leading to delayed fluorescence. Coordination polymers are intricate systems comprising metal ions, one or more types of ligands, and the potential inclusion of solvent molecules or other guests within voids. These polymers can emit light due to different electron transitions, including intraligand (ligand-centered), metal-centered, metal-to-ligand, and ligand-to-metal charge transfer (MLCT and LMCT) transitions. Additionally, electron transitions occurring in guest molecules enclosed within the pores of coordination polymers can impact their photophysical characteristics [16].

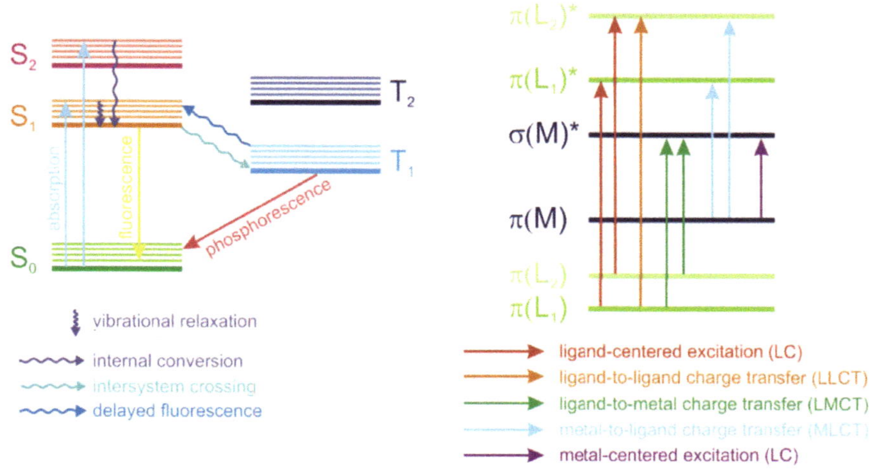

Figure 7.2 The left side shows a Jablonski diagram depicting the processes involving absorption and emission of radiation, while the right side provides a schematic representation of energy levels within coordination polymers (Obtained from open access paper) [38].

Typically, the excited electronic states of the CT complex and exciplex result from a combination of CT states and exciton resonance contributions originating from locally excited states among their constituents. The functions associated with the wavelength can be expressed as:

$$\Psi_1^*(CT)^* \quad C_1\psi(AD)^* \quad C_2\psi(A^+D^-)^* \tag{7.1}$$

$$\Psi_2^*(EX)^* \quad C_{1'}\psi(AD)^* \quad C_{2'}\psi(A^+D^-)^* \tag{7.2}$$

$$\Psi_3^*(CS)^* \quad \psi(A^{''+} + D^{''-})^* \tag{7.3}$$

In accordance with Kitagawa et al. study, configurations involving face-to-face arrangements (π-stacking) between the electron-A of naphthalenediimide (NDI) and the electron-D aromatic guests on site A (characterized by a slit-type pore where aromatic guest molecules interact in a face-to-face orientation with the NDI ligand) enable effective orbital overlap, leading to a greater covalent contribution in the excited state ($C_{2'} > C_2$ in Eqs. (7.1 and 7.2)). These configurations are referred to as exciplex ($(A - D)^*_{EX}$). In contrast, the arrangement of aromatic guests situated at site B (featuring an undulating one-dimensional channel where aromatic guest molecules have less structural confinement) in relation to the NDI pillar is more open. Consequently, the NDI–guest complex exhibits a more distinctly ionic character ($C_2 > C_{2'}$ in Eqs. (7.1 and 7.2)) and is termed a CT complex ($(A^{\delta-} - D^{\delta+})^*_{CT}$), where *CT* represents charge transfer, *EX* stands for exciplex, and *CS* denotes charge separation. It is well-established that a slight increase in solvent polarity typically results in a significant reduction of $(A - D)^*_{EX}$ emission, promoting the transition to the excited CT state ($(A^{\delta-} - D^{\delta+})^*_{CT}$). This state is characterized by lower emission efficiency (mostly nonfluorescent in liquid solutions) and a larger red shift compared to the exciplex. The distinctive adsorption sites within porous coordination polymer facilitated the identification of two discernible complexes (exciplex and CT complex). This distinction arose from the varied conformations adopted by the aromatic guests in relation to the NDI moiety. Additionally, akin to the influence of solvent polarity on the fluorescence of the exciplex and CT complex in solution, the emission in this porous coordination polymer was found to be more favored for the CT complex over the exciplex when the guest exhibited a higher dipole moment and/or stronger electron-donating capability. The porous coordination polymer featuring guests with robust electron-donating properties could promote the entire electron transfer process, leading to fluorescence quenching through the conversion to radical-ion pair $(A^{\bullet-}D^{\bullet+})_{RIP}$ states. Accordingly, porous coordination polymer with an anisole molecule, a guest with a relatively high dipole moment and strong electron-donating ability, showed a much reduced quantum yield, $\Phi_{fl} \approx 0.01$, and the long lifetime value of 6.6 ns is much shorter than τ4 attributed to $(A^{\delta-} - D^{\delta+})^*_{CT}$ states, denoting a partial deactivation process to the $(A^{\bullet-}D^{\bullet+})_{RIP}$ state [17]. In addition, Maji et al. presented a flexible 3D supramolecular framework {[Zn(ndc)(o-phen)]·DMF}n (o-phen=1,10-phenanthroline, ndc=2,6-napthalenedicarboxylate) featuring confined nanospaces capable of hosting various electron-donating aromatic amine guests, leading to selective turn-on emission signals. This framework serves as a molecular recognition platform, employing an emission-readout process. The ability to achieve unprecedented tunable emission with different amines is ascribed to the formation of emissive CT complexes with o-phen linkers. In certain cases, this CT emission is further amplified by energy transfer from the chromophoric linker unit ndc [18]. Furthermore, Sugimoto et al. synthesized the decoding host by incorporating NDI into the scaffold of an entangled porous framework, characterized by structural dynamics due to the dislocation of two chemically non-interconnected frameworks. The wide range of emission stems from the enhanced NDI–volatile organic compounds (VOC) interaction, resulting in either exciplex fluorescence with CT characteristics or phosphorescence stabilized by the heavy atom effect. These emission wavelengths not only exhibit reasonable separation but also have intensities amplified enough to be discernible by the naked eye, whereas the framework without VOCs remains non-emissive. Leveraging this turn-on switching property, these sensors demonstrate a nonlinear luminescence response to VOC vapor concentration, with higher sensitivity in the lower-vapor-pressure range of VOCs. The distinctive photophysical properties of NDI enable the application of a strategic decoding protocol. NDI interacts with aromatic VOCs, exhibiting an exciplex emission modulated by a host–guest CT interaction. This modulation in emission wavelength, and consequently emission color, serves as a readout corresponding to the chemical substituent on the phenyl ring. While numerous studies have reported the synthesis and structure of PCPs with NDI derivatives, the photochemical properties of these compounds remain undescribed. This gap can be attributed to the inherently weak fluorescent properties of NDI [19]. Moreover, Maji et al. studied an unprecedented reversible electron transfer phenomenon within a redox-active MOF, with a particular focus on understanding the reversible control of the two electronic states. The deliberate design of organic chromophores (linkers) and metal nodes or secondary building units (SBU) in a coordination framework allows for precise management of the spatial orientation of chromophores. The approach involves employing anthracene dicarboxylate (adc) and N,N'-di(4-pyridyl)-1,4,5,8-naphthalenediimide (bpNDI) as the D-A pair within a pillared-layer type MOF structure. Upon the removal or encapsulation of guest solvent, this D-A MOF demonstrates reversible electron transfer while maintaining overall structural order. In contrast, a control isostructural MOF, incorporating 1,2-bis(4-pyridyl)ethylene (bpee) as a pillar via the solvent-assisted linker exchange (SALE) method, does not exhibit ET, confirming the essential role of bpNDI as the key electron-A component in the D-A MOF 1. Additionally, spectroscopic, and theoretical analyses elucidate the involvement of paddlewheel-type SBUs in the reversible transformation of the charge-separated state [20].

The selection and arrangement of D and A are crucial factors that determine the electronic properties of a CT complex. In organic semiconductors, the bandgap typically corresponds to the energy difference between the highest occupied molecular orbital (HOMO) of D and the lowest unoccupied molecular orbital (LUMO) of A. The bandgaps in D-A polymers can be continuously modulated by combining various functionalities and diverse fragments with different electrical characteristics. This allows for precise control and alteration of the bandgap to meet specific requirements and achieve desired electrical properties [21]. The use of flexible, lightweight, and transparent D-A polymers, along with a straightforward and cost-effective solution-based film fabrication process, enables large-scale production capabilities [22]. However, D-A small molecules or polymers often face challenges such as electron-hole recombination due to trap sites, low charge carrier mobilities, and disordered accumulation of molecules and polymers. To address these issues, preventing electron-hole pair recombination and enhancing charge carrier transport can be achieved through long-range ordered arrangements and integration of D and A moieties. The self-organization of D-A components also facilitates CT by creating a bicontinuous interpenetrating network with significant interfacial regions. Advancing this technology requires extending the spectral response and controlling the morphology at the atomic or molecular level. However, achieving fine morphological control at this level in such materials can be difficult due to the need to adjust various operating factors, including solvent viscosity [23], blend composition [24], thermal annealing conditions [25], and additive levels [26]. These manipulations present synthetic challenges and involve complex implementation procedures [27]. In general, flexible side chains are frequently introduced into molecular architectures to improve the solubility of D-A polymers in organic solvents. These chains consider the strong π-π interactions between the linear conjugated backbones. Introducing flexible side chains enables D-A polymers to achieve improved solubility and better compatibility with organic solvents [28]. Furthermore, it is worth mentioning that the interaction of halogenated compounds with aromatics leads to the development of complexes that exhibit intermolecular charge transfer (ICT). The first organic conductor, [TTF]Cl, was discovered because of this phenomenon [29]. The ICT between D and A moieties has generated significant interest in the fields of materials science and chemistry. This interest was driven by the discovery of organic salts, such as tetrathiafulvalene-tetracyanoquinodimethane (TTF-TCNQ) [30]. Furthermore, organic small molecules incorporating inorganic positive-negative (P-N) junctions have exhibited a rectification effect with covalently bonded D-A pairs under external electric fields [31]. This phenomenon is similar to the effects of rectification frequently observed in inorganic P-N junctions [32]. In organic semiconductors like COFs, efficient charge transport relies on the presence of an electronic coupling element, commonly referred to as the CT integral (t). This coupling element facilitates CT between closely spaced π-conjugated building blocks [33]. The CT integrals in organic semiconductors are relatively significant, approximately 0.1 eV, while intrinsic carrier mobilities often range from 0.01 to 1.00 $cm^2\ V^{-1}\ s^{-1}$. The intermolecular distance has an exponential effect on the spatial overlap between frontier orbitals, which determines t. In contrast, MOF linkers have less electrical coupling since they are normally positioned far apart, and their frontier orbitals rarely interact. MOFs have poor charge transport properties because binding to ions considerably increases the activation energy for CT [34]. Two main CT mechanisms have been reported in conductive polymers. In electronically delocalized polymers like polypyrrole and polythiophene, a highly conjugated bond system plays a role in the first process. The motion of electrons along the conjugated backbone in these systems facilitates CT. The second process involves site-to-site electron hopping in polymers that contain discrete localized redox sites like ferrocenes and pyridyl metal complexes, together with the diffusion of counterions to maintain charge balance. This mechanism is particularly relevant in describing CT processes through MOFs, where the localized redox sites play a crucial role in facilitating CT [35]. It should be mentioned that the processes of CT in MOFs remained poorly understood even after approximately 20 years of extensive research. The exploration, validation, and modeling of new strategies to enhance charge mobility in MOFs remain difficult and challenging [36]. Moreover, a partial CT is performed when the LUMO of the A molecule and the HOMO of the D molecule are electronically connected. This results in a partly ionic ground state, designated as $D^{\rho+}A^{\rho-}$. The degree of CT (ρ) is responsible for the transport properties observed, and it is determined by the ionization potential of the D (I_D) and the electron affinity of the A (E_A). When the combination of D-A results in a large difference between I_D and E_A, much greater than the electrostatic Madelung energy (EM), a quasi-neutral CT state with a ρ less than 0.5 is produced. On the other hand, when the I_D and E_A difference is much smaller than EM, a much higher ρ occurs, resulting in a quasi-ionic (D^+A^-) or ionic CT [37]. Considering the report on MOFs by Miyasaka, the transition from a D to an A, referred to D^0A^0 to $D^{\delta+}A^{\delta-}$, induces changes in charge distributions, leading to the formation of a set of $S = 1/2$ spins with exchange interactions and a structural modification in the ionic form $D^{\delta+}A^{\delta-}$. These modifications have an immediate effect on the macroscopic magnetic ordering, dielectric responses, and electron and hole transport physical characteristics. Consequently, the design of CT systems offers enormous potential for discovering novel functional materials, including hybrid materials with magnetic and conductive properties. However, the majority of D-A-built systems are currently restricted to supramolecular assemblies based on organic molecules with π-stacking motifs, which only produce a triplet state. This behavior resulted due to the ferromagnetic coupling of the generated $S = 1/2$ spins. Utilizing metal complexes such as D and/or A provides notable benefits over organic molecules, particularly in terms of a diverse selection of redox-active metal complexes. The coordination bonding of metal complexes allows for a direct connection between the D and A moieties [35c]. Additionally, many metal complexes exhibit stereospecific properties that enable exact control of the lattice dimensions during D-A MOF assembly. Furthermore, the static multiple spin state in d-orbitals can be used for purposes other than electron transfer. This creates opportunities for several CT mechanisms involving d and p-orbital electrons, such as metal-to-metal, metal-to-ligand, ligand-to-ligand, and ligand-to-ligand CT. However, finding an appropriate D-A combination might be difficult because CT must be properly controlled by fine-tuning the EA and IP. This increases the challenge of developing functional molecular systems [38].

Theory of CT: The rate of hole-electron transfer in intermolecular structures ($K_{h/e}$), such as the adjacent linkers in MOFs with a weakly coupled supramolecular system, can be described by the Marcus Equation (Eq. (7.4)).

$$K_{h/e} = \frac{2\pi}{h} t^2_{h/e} \frac{1}{\sqrt{4\pi\lambda_{h/e} K_b T}} \exp\left[-\frac{\left(\lambda_{h/e} + \Delta G^0\right)^2}{4\lambda_{h/e} K_b T} \right] \tag{7.4}$$

where t is the CT integral that represents the electronic coupling between the initial and final states. It quantifies the strength of the interaction involved in CT.

$$t_{h/e} = \Psi_{D^{\pm}A^0} | V_{DA} | \Psi_{D^0 A^{\pm}} \tag{7.5}$$

where, the initial and final states of the hole-electron localization on A and D molecules are represented by the wavefunctions $\Psi_{D^0 A^{\pm}}$ and $\Psi_{D^{\pm}A^0}$ respectively. λ indicates the energy needed for the system to modify its solvent and bonding conditions before the transfer of electrons to the final state is designated as the reorganization energy. λ can be further decomposed into internal $\lambda_{h/e}^{int}$ and solvent $\lambda_{h/e}^{solv}$ energy terms. The definitions of $\lambda_{h/e}^{int}$ and $\lambda_{h/e}^{solv}$ are as follows:

$$\lambda_{h/e}^{int} = \left[E^{\mp}\left(g^{0}\right) - E^{\mp}\left(g^{\pm}\right)\right] + \left[E^{0}\left(g^{\pm}\right) - E^{0}\left(g^{0}\right)\right] \quad (7.6)$$

$$\lambda_{h/e}^{solv} = \frac{e^2}{8\pi}\left(\frac{1}{\varepsilon_\infty} - \frac{1}{\varepsilon_0}\right)\left(\frac{1}{2R_D} + \frac{1}{2R_A} - \frac{1}{R_{AD}}\right) \quad (7.7)$$

where, $E^{\pm}$ (g^0) represents the energy of the monomer with an excess charge q = ±1, while the nuclear geometry g^0 corresponds to the equilibrium state of the neutral monomer unit. R_A refers to the inter-linker distance and R_{AD} represents the radius of the π-electron cloud associated with the excess charge on the D-A. Additionally, ΔG^0 denotes the energy difference between $\Psi_{D^0A^{\mp}}$ and $\Psi_{D^{\pm}A^0}$.

If the electronic coupling between the D and A states is insignificant, the CT can occur through the virtual states of bridging molecules, employing a super-exchange mechanism. In this case, Eq. (7.5) can be used.

$$t_{eff} = {t_{DB}t_{BA}}\big/{\left(E_D - E_B\right)} \quad (7.8)$$

where $E_D - E_B$ represents the energy difference between the node and linker states. Similarly, t_{DB} and t_{BA} denote the CT integrals between the D-node/bridging-linker and bridging linker/A-node states, respectively [34c]. Therefore, scientists have realized that creating CT systems is a successful strategy for investigating novel functional materials, such as magnetic and conductive hybrids, which exhibit synergistic features [38].

7.3 Synthetic Porous Frameworks

The unique characteristics of synthetic porous frameworks, including pore size, physical and chemical stability, synthesis methods, and extended π-conjugation (organic frameworks), have a significant impact on their final properties [39]. Among these synthetic porous materials, CMPs, COFs, and MOFs have been studied for their potential in CT applications such as optoelectronics, energy storage [40], separation and storage [41], chemical sensors [42], and photocatalysis [43]. CMPs are a type of 3D network-structured material that is distinguished by its π-conjugated system. According to the International Union of Pure and Applied Chemistry (IUPAC), CMPs are materials with a permanent pore size smaller than 2 nm [44]. COFs, on the other hand, are porous, stable, and crystalline structures formed through the reaction between organic precursors. This reaction results in the formation of strong covalent bonds, leading to the arrangement of these bonds in 2D or 3D configurations [45]. Furthermore, MOFs are a unique class of materials that combine extended metal ion and organic ligand interconnection with a permanently meso-microporous skeleton. The wide range of molecular building blocks that can be connected in this manner offers the ability to manipulate functionality and structure in porous materials. Since these porous frameworks have been explored for CT, they have garnered significant attention due to their varied structural landscapes. This study will survey the progress of research on CMPs, COFs, and MOFs in CT applications and provide a perspective on the future of these materials.

7.3.1 CT Based on CMPs

CMPs represent a category of porous polymeric materials in which the backbones are interconnected through extended π-conjugated units. Typically, CMP precursor molecules consist of a central core and building blocks that are later linked by covalent bonds to form a 3D structure with π-conjugation [44b]. The incorporation of electro-reactive groups, such as pyrroles, thiophenes, carbazoles (Czs), and similar moieties, as side groups enables the formation of a 3D framework with a microporous-mesoporous structure [46]. Based on the research in this field, covalent triazine frameworks (CTFs) [47] are well-known as a subclass of CMPs because they have extended π-conjugation with microporous structure [48]. Recently, Faul et al. achieved CMPs with 20 times higher surface area than the standard reaction by utilizing the Buchwald-Hartwig cross-coupling reaction with different-sized cations or anions [49]. The rigid and interlinking architecture of CMPs restrains the rotation of the benzene ring in these structures, thereby preventing the quenching of fluorescent groups [50]. Additionally, the introduction of various functional groups into CMPs facilitates their utilization as luminescent materials [44b, 51]. Therefore, the advantages of CT based on CMPs can be described as follows. CMPs with tunable pore structures, compositions, and high surface areas offer an effective design strategy for gas storage and separation [52]. CMPs can be utilized as successful photocatalytic materials due to their extensive π-conjugation and tunable bandgap. The large specific surface area and fine pores of CMPs improve the performance of sensors by enabling enhanced analyte contact and penetration. This characteristic enhances sensing capabilities and sensitivity in CMP-based sensor systems [42]. CMPs possess advantageous properties in electrochemical processes and energy storage. Their micropore sizes facilitate rapid transport of electrons and ions between the electrode and electrolyte, leading to efficient CT, enhanced electrochemical reactions, and improved overall performance in CMP-based electrochemical devices and energy storage systems [44b, 53].

Figure 7.3 illustrates the skeletal configurations of various D-A complexes utilizing CMPs. Here, some D-A units are mentioned based on CMPs. El-Mahdy et al. synthesized D-A CMPs—Cz-Py, Cz-TPT, Cz-BT, and Cz-TzTz semiconducting CMPs for visible-light-driven H_2 and O_2 evolution through water splitting. This D-A CMP was created through 3,6-bis(4,4,5,5-tetramethyl-1,3,2-dioxaborolan-2-yl); 9-(4-(4,4,5,5-tetramethyl-1,3,2-dioxaborolan-2-yl) phenyl); 9H-Cz (Cz-3BO) to provide the D moiety, and 1,3,6,8-tetrabromopyrene (Py-4Br) [54], 2,4,6-tris(4-bromophenyl); 1,3,5-triazine (TPT-3Br) [55], 4,7-dibromobenzo[c] [56], thiadiazole (BT-2Br) [57], and 2,5-bis(4-bromophenyl)thiazolo[5,4-d] thiazole (TzTz-2Br) to provide the A units (Figure 7.3a) [58]. In another study, by simply doping TCNQ in dichloromethane at room temperature, Nagai et al. achieved the CT complexation of SBF-Py-CMP with equimolar quantities of pyrene (Py) and TCNQ (SBF-Py-CMP:TCNQ = 1:1) (Figure 7.3b) [2]. Furthermore, BCMP-3 was synthesized by Liu et al. through a Suzuki-Miyaura cross-coupling polycondensation reaction. The reaction was carried out in an alkaline environment with Pd(0) as the catalyst by tris(4-bromo-2,6-dimethylphenyl) borane and tris(4-dihydroxyboranylphenyl)amine (TBPA) as the monomers (Figure 7.3) [59]. In another study, Jiang et al. successfully produced PP-CMP using the Suzuki cross-coupling reaction between 1,2,4,5-tetrabromobenzene (TBB) and benzene-1,4-diboronic acid (BDBA). Plus, the TPE-CMP was produced by the Suzuki reaction between BDBA and tetrakis(4-bromophenyl)ethene (TBTPE)

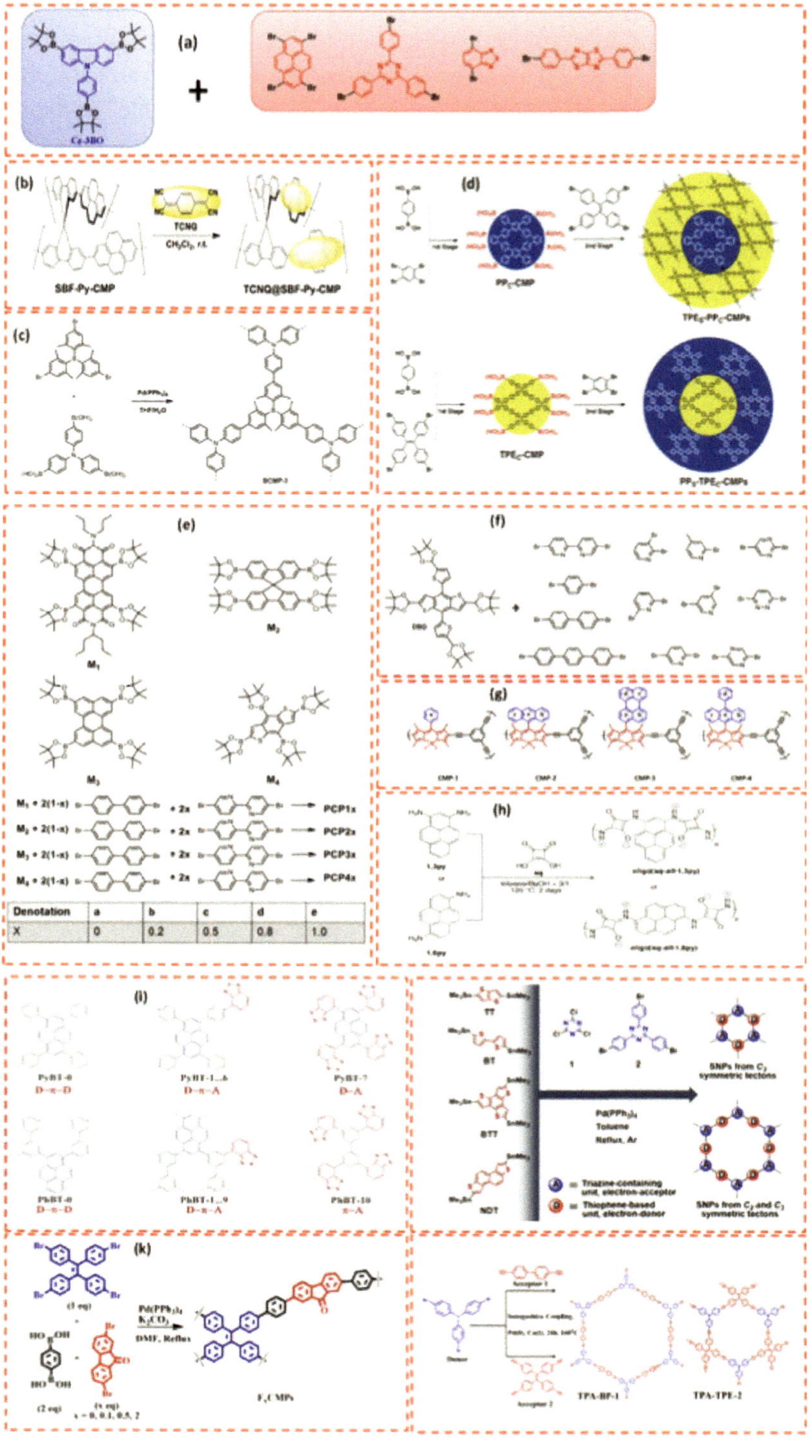
(a)
(b)
TCNQ
SBF-Py-CMP
TCNQ@SBF-Py-CMP
(c)
(d)
PP$_C$-CMP
TPE$_S$-PP$_C$-CMPs
TPE$_C$-CMP
PP$_S$-TPE$_C$-CMPs
(e)
M$_1$
M$_2$
M$_3$
M$_4$
PCP1x
PCP2x
PCP3x
PCP4x
Denotation a b c d e
X 0 0.2 0.5 0.8 1.0
(f)
(g)
(h)
(i)
D–π–D
D–π–A
D–A
π–A
TT
BT
BTT
NDT
Pd(PPh$_3$)$_4$
Toluene
Reflux, Ar
SNPs from C_3 symmetric tectons
SNPs from C_2 and C_3 symmetric tectons
(k)
Pd(PPh$_3$)$_4$
K$_2$CO$_3$
DMF, Reflux
TPA-BF-1
TPA-TPE-2

Figure 7.3 Skeletal structures of several CMPs-based D-A (a) D-A CMPs—Cz-Py, Cz-TPT, Cz-BT, and Cz-TzTz CMPs (Reprinted from [58] with the permission of Elsevier). (b) SBF-Py-CMP with TCNQ CT complex (Reprinted from [2] with the permission of Wiley). (c) BCMP-3 synthesis (Reprinted from [59] with the permission of Wiley). (d) TPES-PPC-CMPs and PPS-TPEC-CMPs (Reprinted from [51] with the permission of RSC). (e) monomers and synthesis of PCP (Obtained from ACS publications) [60]. (f) PCP with monomers DBD (Obtained from ACS publications) [61]. (g) PAH-D and Bodipy-A (Reprinted from [62] with the permission of Wiley). (h) squaric acid and X-diaminopyrenes with TCNQ (Reprinted from [63] with the permission of RSC) (i) D-π-A CMPs (Py as D and benzothiadiazole as A unite) (Reprinted from [64] with the permission of Elsevier). (j) thiophene-based D (in red) with triazine containing A (in blue) (Reprinted from [65] with the permission of Wiley). (k) TPE as D and 9-fluorenone as A (Reprinted from [66] with the permission of Wiley). (l) tris(4bromophenyl) amine with 1,1,2,2,–4,4′-diethynylbiphenyl as D and tetrakis(4- ethynylphenyl) ethane as A (Reprinted from [67] with the permission of RSC).

(Figure 7.3d) [51]. Furthermore, the PCPs were synthesized through palladium-mediated polycondensation by Suzuki polycondensation, involving the copolymerization of several chromophore monomers with biphenyl (bph) and bipyridyl (bpy) co-monomers (Figure 7.3e) [59]. Also, the preparation of PCP photocatalysts was carried out by Yu et al. through palladium-mediated Suzuki polycondensation using DBD monomers in (N,N-dimethylformamide) DMF at 130 °C for 24 hours (Figure 7.3f) [61]. Furthermore, Zhang et al. improved the photosensitization of the original CMP (CMP-1) by adding polycyclic aromatic hydrocarbons (PAHs) generated from diiodo-Bodipy derivatives and 1,3,5-triethynylbenzene (TEB) via a Sonogashira coupling process. This synthetic method led to the development of several CMPs (CMP-2-4) (Figure 7.3g) [62]. Plus, Nagai showed how to create extremely crystalline π-conjugated oligomers through a simple solvothermal reaction by diaminopyrenes and squaric acid. Oligo(sq-alt-1,3py) and oligo(sq-alt-1,6py) were produced by preferentially linking the resultant oligomers at the 1,3- and 1,6-positions of the Py units, respectively (Figure 7.3h) [63]. Moreover, Jiang et al. examined CMPs-based D-A photocatalysts with exceptional photocatalytic activity for hydrogen evolution. They used Py with a planar structure and prolonged π-conjugation, as the D unit and benzothiadiazole as the A unit. Furthermore, benzene or biphenyl was considered the crosslinker unit in the structures of the CMPs (Figure 7.3i) [64]. In addition, the sulfur and nitrogen-containing polymers (SNPs), a set of eight porous polymers with comparable geometries, were successfully synthesized by researchers. These polymers were designed by combining electron acceptors triazine-(C_3N_3) and thiophene containing tectons as electron donors, namely bithiophene (BT), thienothiophene (TT), naphthodithiophene (NDT), and benzotrithiophene (BTT). These covalently-bonded D-A frameworks demonstrated permanent porosity, making them highly promising porous materials (Figure 7.3j) [65]. Additionally, the synthesis of several CMPs, including F0.1CMP, F0.5CMP, and F2.0CMP, was carried out by researchers. TPE and 9-fluorenone (F) were used to create these CMPs as the D and A units, respectively. By engineering the bandgap of these CMPs, the optical properties were fine-tuned to enhance their photocatalytic performance in water splitting for hydrogen production (Figure 7.3k) [66]. Moreover, Maji et al. synthesized electrochemically active CMPs, namely TPA-BP-1 and TPATPE-2. These CMPs were prepared through the Sonogashira–Hagihara coupling reaction, combining tris(4-bromophenyl)amine as the D unit with 1,1,2,2,–4,4'-diethynylbiphenyl and tetrakis(4-ethynylphenyl)ethane as the A units, respectively (Figure 7.3l) [67].

Here, some recent studies on CT complexes through CMPs are reported with more details summarized in Table 7.1. Nagai et al. presented one of the initial studies on the synthesis of CT-complexed CMPs with excellent porosity and thermal stability. Their research focused on designing and synthesizing a CT complex using CMP as the D component and TCNQ as the A unit. Specifically, they developed a D-A complex called TCNQ-SBF-Py-CMP, which involved incorporating spirobifluorene (SBF-Br) and Py (diboronic acid bis(pinacol) ester (Py-Bpin)) units in an alternating order. The resulting CMPs exhibited nonextended π-delocalization in the solid state, distinguishing them from linear conjugated polymers. The 3D CT-complexed CMPs produced by the rapid synthesis showed chemical and thermal stability (up to 210 °C) with a high surface area (400 m^2 g^{-1}) and precise control over the concentration of D-A pairs [2]. In addition, Nagai et al. in their previous research showed the synthesis of π-conjugated oligomers containing CT complexes achieved in a single step using the solvothermal reaction of diaminopyrenes and squaric acid. The synthesis led to the formation of oligomers bonded at the 1,3- and 1,6-positions of the Py units, resulting in oligo(sq-alt-1,3py) and oligo(sq-alt-1,6py). These oligomers indicated an interesting mechanochromic property, changing from orange to a rich metallic green. Furthermore, the presence of intramolecular CT revealed their potential application in various phenomena such as spin density waves (SDW), charge ordering, spin-Peierls ground state, and superconductivity. Coronene as an aromatic molecule demonstrated π-electron systems capable of donating electrons based on their ionization potential (I). When TCNQ is in contact with electron donors, it is quickly reduced to the open shell electron radical anion TCNQ [63]. Photocatalytic hydrogen evolution from water has been extensively investigated in various studies based on CMPs. Maji et al. investigated the design and synthesis of the D-A conjugated polymers (FxCMP) composed of TPE as the D unit and 9-fluorenone (F) as the A unit. By adjusting the D-A ratio, the efficiency of energy transfer was controlled, resulting in a range of tunable emission colors from green to red. As the results, F0.5CMP with the optimum bandgap of 2.3 eV, demonstrated the highest hydrogen evolution rate (HER) of 659.55 μmol g^{-1} h^{-1} under visible light irradiation with a thermal stability of 600 °C, and surface area of 537 m^2 g^{-1} [68]. In addition, El-Mahdy et al. presented a study on semiconducting D-A CMPs as photocatalysts for hydrogen (H_2) and oxygen (O_2) evolution through visible light-driven water splitting. Pyrene (Py), triphenyltriazine (TPT), benzothiadiazole, and thiazolylthiazole (TzTz) were used as A units, and employed a Cz moiety as the D unit. The outcomes showed that the D-A CMPs had strong thermal stabilities (Td10: up to 623 °C; char yield: up to 83 wt%) and large surface areas up to 1530 m^2 g^{-1}. Additionally, the Cz-TzTz CMP demonstrated a good HER of 15.3 μmol g^{-1} h^{-1} without the need for an additional Pt co-catalyst by employing ascorbic acid as the sacrificial electron-D. Moreover, the CMPs exhibited an oxygen evolution rate (OER) of 3.38 μmol g^{-1} h^{-1} in the presence of $AgNO_3$ as an electron-A, making them promising photocatalysts for solar energy harvesting and conversion [58]. Furthermore, Zhou et al. created a CMP film as hole-carrying layers in OPVs by in-situ electrochemical deposition. The CMP Cz groups have shown the ability to chemically couple at a low oxidation potential, resulting in an extended π-conjugated backbone. This characteristic was beneficial for facilitating CT along the main chain of the polymer. The PTB7-Th:PC71BM solar cell showed slightly better performance with the CMP HTL compared to the widely used PEDOT:PSS led to an increase in both V_{oc} and J_{sc} [46]. Moreover, dithienothiophene (DTT)-based CMPs were examined utilizing bulk and electrochemical oxidative polymerizations by Thomas et al. The spectro-electrochemical measurements revealed that DTT-CMP exhibited significant changes in absorption properties upon reversible oxidation and reduction, making it a promising material for electrochromic devices. Additionally, DTT-CMPs showed potential for applications in organo(opto)electronic devices and sensors due to their ability to undergo redox switching and incorporate electron-donating DTT units into a conjugated porous network [69]. Furthermore, Zhang et al. reported the synthesis of CMPs (CMP-2–4) by incorporating polycyclic PAHs into the parent CMP-1 to enhance photosensitization capabilities. The results revealed that CMP-2 effectively promoted methyl phenyl sulfoxide photosynthesis with a yield of 92% with a surface area of 478.6 $m^2\ g^{-1}$. The separation of electron-hole pairs in the CMPs was finely tuned by adjusting the form and polymerization number of aromatic rings, further enhancing their performance [62]. In CMPs photoactive studies, Jiang et al. focused on investigating CMPs utilizing two commonly used thiophene structures, namely 1,3,5-tri(2-thienyl)-benzene (TTB) and BTT attached to a central aromatic benzene moiety. The creation of porous π-skeletons was achieved through the alignment of multiple arms and was predicted by the computed optimal elementary one-pore structures. The findings confirmed that these semiconducting CMPs can be used as photoactive layers for efficient photoenergy conversion. The specific monomer structures played a crucial role in controlling the electronic properties of the CMPs and guiding the growth of porous skeletons, thereby influencing their conductivity and photovoltaic characteristics. In this case, C_{60} was selected as the electron-A with a diameter of 0.7 nm, which was smaller than the pore sizes of the CMP films [24, 70]. Plus, studies have illustrated that D-π-A polymers significantly improve photocatalytic performance compared to D-π-D polymers without an A unit or D-A polymers without a π crosslinker unit. In this area, Jiang et al. synthesized D-π-A CMP photocatalysts using benzene (biphenyl), Py, and benzothiadiazole as π crosslinker, D, and A units, respectively. Remarkable photocatalytic HER up to 106 $\mu mol\ h^{-1}$ was obtained on PyBT-2, with a Py to benzothiadiazole ratio of 9:2, under UV-visible light irradiation ($\lambda > 300$ nm). The PyBT-2 structure showed high thermal stability up to 500 °C and a surface area of 847 $m^2\ g^{-1}$. According to the research, the photocatalytic performance of the resulting CMP photocatalysts was found to be significantly influenced by the ratio of D to A [64].

Table 7.1 Recent studies based on CMPs for CT (D-A)

CT-CMPs	Surface area m^2g^{-1}	Thermal stability (°C)	Bandgap energy (eV)	CT performance	Ref.
25%TCNQ@ SBF-Py-CMP	400	210	----	Strong emission band in the 400–548 nm region	[2]
Cz-Py CMP	1530	587	2.38	HER = 3761 μmol g^{-1} h^{-1}	[58]
Cz-TPT CMP	377	499	2.34	HER = 1780 μmol g^{-1} h^{-1}	[58]
Cz-BT CMP	300	623	1.99	HER = 416 μmol g^{-1} h^{-1}	[58]
Cz-TzTz CMP	104	498	2.10	HER = 15315 μmol g^{-1} h^{-1}	[58]
CMP-1 (PAHs=D, Bodipy=A)	354.5	-----	----	Yield of photosynthesis of methyl phenyl sulfoxide = 32 %	[62]
CMP-2 (PAHs=D, Bodipy=A)	478.6	-----	----	Yield of photosynthesis of methyl phenyl sulfoxide = 92 % yield	[62]
CMP-3 (PAHs=D, Bodipy=A)	316.7	-----	----	Photosynthesis of methyl phenyl sulfoxide	[62]
CMP-4 (PAHs=D, Bodipy=A)	604.9	-----	----	Photosynthesis of methyl phenyl sulfoxide	[62]
1,3: oligo(sq-alt-1,3py)-TCNQ	-----	-----	----	After mechanical grinding, the Absorption color was noticeably changed from orange to a metallic deep green tone	[63]
1,6: oligo(sq-alt-1,6py)-TCNQ	-----	-----	----		[63]

(*Continued*)

Table 7.1 (*Continued*)

CT-CMPs	Surface area m^2g^{-1}	Thermal stability (°C)	Bandgap energy (eV)	CT performance	Ref.
PyBT-2	865	500	2.21	HER= 106.2 μmol g^{-1} h^{-1}	[64]
PhBT-4	930	500	2.17	HER= 85.2 μmol g^{-1} h^{-1}	[64]
$F_{0.5}$CMP	537	600	2.3	HER= 659.55 μmol g^{-1} h^{-1}	[68]
TPA-BP-1	468	-----	2.19	Semi-conduction and redox-activity, appreciable ORR activity, and catalytic stability	[67]
TPA-TPE-2	117	-----	1.69	Semi-conduction and redox-activity, appreciable ORR activity, and catalytic stability	[67]
PCP-DBD	279	-----	1.81	Hydrogen production rate at ~33 μmol h^{-1}	[60]
CMP film	----	----	----	Hole transporting layers in OPVs	[46]
DTT-CMP	790	----	----	Electrochromic devices, organo(opto) electronic devices and sensors	[69]
BTT-CMP-C_{60}	BTT-CMP = 1152	----	1.91	Applications in photocatalysis, energy storage, and optoelectronics	[71]
TTB-CMP-C_{60}	TTB-CMP = 1405	----	2.18	Highly efficient photoactive layers for energy conversion	[72]

7.3.2 CT Based on COFs

COFs are recognized as a specific category of porous materials characterized by the arrangement of organic building blocks connected through covalent bonds, resulting in a crystalline structure. Therefore, the advantages of CT based on COFs can be described as follows. COFs are highly tunable, allowing precise control over their properties. The modular nature of COFs facilitates the design of materials with specific functionalities. COFs often possess conjugated structures, which can contribute to enhanced charge transport properties. Some COFs are solution-processable, enabling their integration into thin films or coatings for electronic devices. However, some COFs may exhibit limited stability, particularly in the presence of certain solvents or under harsh conditions. The groups of Jiang [73], and Wang [3a], exhibited early investigations into the utilization of different commonly used ligands as "linkers" and porphyrin as the "core" for the synthesis of 3D-COFs. Subsequently, other research groups have made significant contributions to the development of porphyrin-based 3D-COFs [24, 27, 74]. In comparison to their 3D counterparts, 2D-COFs are typically constructed using nonplanar parts with electronically localized nodes. This unique structural feature provides them with distinct advantages in forming ordered π-arrays for CT applications [74].

Figure 7.4 illustrates the skeletal configurations of various D-A complexes utilizing COFs. Here, some D-A units are mentioned based on COFs. Gu et al. reported the synthesis of TTF-COF1, created through imine condensation of tetraformyl-tetrathiafulvalene (TTF-fo) and 2,6-diaminoanthraquinone (DAQ) as electron-D and A, respectively (Figure 7.4a) [75]. In addition, Bein et al. prepared a PCBM:TT-COF complex by incorporating PCBM into the TT-COF. Thieno[3,2-b]thiophene-2,5-diyldiboronic acid (TTBA) and the polyol 2,3,6,7,10,11-hexahydroxytriphenylene (HHTP) were combined to form the TT-COF under solvothermal conditions (Figure 7.4b) [76]. Furthermore, Liu et al. prepared two D-A COFs through BTT and TPT/triphenyltriphenyl in the skeleton as D and A, respectively. The appropriate PSCs were created once the D-A COFs were incorporated into the FAPbI3 layer (Figure 7.4c) [77]. Also, in a comprehensive study, periodic and vertical D-on-D and A-on-A columnar arrays were synthesized by Jiang et al. The triphenylene D units located at the vertices and the benzothiadiazole A units on the edges created 2D hexagonal polygons at the monolayer level, with all atoms on a single plane (Figure 7.4d) [78]. Moreover, Jiang et al. prepared a hexagonal triphenylene–diimide D–A COF. The D-A COFs exhibited a layer structure involving π-stacking led to the formation of periodic arrangements of triphenylene-on-triphenylene and diimide-on-diimide columns. The spatial arrangement between these columns confirmed that photochemical processes occur predominantly within the neighboring columns, allowing for controlled and localized photochemical events (Figure 7.4e) [79]. In another study, Bein et al. developed a model system consisting of interconnected networks of electron-D and A phases. The structure was constructed using stacked [6,6]-phenyl-C61-butyric acid methyl ester (PCBM) as the electron-A component, and thieno[2,3-b] thiophene-based building blocks as the electron-D component (TT-COF) (Figure 7.4f) [76]. Furthermore, Jiang et al. synthesized the D-A system CS-COF-C60. They employed C2-symmetric tertbutylpyrene tetraone (PT) and C3-symmetric triphenylene hexamine (TPHA) as building blocks to design wire frameworks and open nanochannels within the structure. The microporous region to facilitate the functionalization of complementary fullerene molecules was enabled by a D-A system called CS-COF-C60. This process allowed the combination of extended π-conjugation with bicontinuous order to occur (Figure 7.4g) [80]. Additionally, Shustova et al. synthesized D-A alignments using 2,5-bis(2-propynyloxy) terephthalaldehyde (BPTA) and 2,5-dimethoxyterephthalaldehyde (DMTA) for coordinative and non-coordinative π-integration, respectively. They also utilized tri-(4-aminophenyl)benzene as a building block (Figure 7.4h) [81]. In a separate study, Shustova et al. incorporated 3D buckyballs as the A component in a host matrix COF. The COF structure was produced by DMTA, tri-(4-aminophenyl) benzene (TAPB), and BPTA materials. The COFs were produced by fluorophore tags via the Sonogashira cross-coupling method (Figure 7.4i) [82]. Furthermore, Jiang et al. reported a D-A based on the COF (DMPc-ADI-COF) D-A with regularly ordered D and A π-columnar arrays created using direct polycondensation processes. In this case, nickel(II), copper(II), and zinc(II) phthalocyanine donors (NiPc, CuPc, and ZnPc) were paired with naphthalene diimide, pyromellitic diimide, and perylene diimide acceptors (NDI, PyrDI, and PDI) (Figure 7.4j) [83]. Plus, Ihee et al. investigated charge carrier dynamics in 2D D-A COFs comprising 5,10,15,20-tetrakis(4-aminophenyl)-21H,23H-porphyrin and 3,4,9,10-perylenetetracarboxylic acid diimide (PDI) (Figure 7.4k) [84]. Also, Jiang et al. presented a comprehensive synthetic approach to transform open lattice COFs into well-organized D-A heterojunctions using a three-component condensation system and the solvothermal approach. They created a succession of intermediate ZnPc-COFs structures that donated electrons. The system was produced by BDBA and 2,5-bis(azidomethyl)-BDBA (N3-BDBA) in various molar ratios as the edges and 2,3-octahydroxyphthalocyaninato)zinc (ZnPc[OH]8) as the vertices. This synthesis approach allowed for the incorporation of different quantities of azide units on the channel walls of the ZnPc-COFs (Figure 7.4l) [85]. Additionally, Nagai and Jiang et al. prepared DZnPc-ANDI-COF composed of zinc phthalocyanine as the electron-D and naphthalene diimide as the A unit. The two units were alternately connected, preserving an electron-transfer distance, and displaying a dihedral angle of about 42 degrees (Figure 7.4m) [86]. Moreover, Jiang et al. presented a strategy involving imine-linked porphyrin COFs, where both fluoro-substituted and non-substituted arenes were incorporated into the edge units in varying molar ratios. Mixtures containing 2,3,5,6-tetrafluoroterephthalaldehyde (TFTA) and terephthalaldehyde (TA) at different molar ratios were employed in the polycondensation with copper 5,10,15,20-tetrakis(p tetraphenylamino) porphyrin (CuP) under solvothermal conditions. This led to the synthesis of five new imine-linked COFs (Figure 7.4n) [87]. Furthermore, Zhang et al. prepared a 2D-COF for organic OPV film using 5-oxo-2,3-dihydro-5H-(1,3)-thiazole-(3,2-a)pyridine-3,7-dicarboxylic acid (TPA) and 5,10,15,20-tetra-(aminophenyl) porphyrin (TAPP) (Figure 7.4o) [27]. In another study, Zhao et al. created CT single-crystals based on D-A interactions. Py and trimeric naphthalene diimide (Tri-NDI) were selected as the electron-D and electron-A, respectively (Figure 7.4p) [1d]. Also, Bein et al. prepared a novel Py-based COF by using 1,3,6,8-tetrakis(4-aminophenyl)Py (1) as its main component. The COF was constructed based on a D-A system and employed an imine-linked rhombic lattice structure known as Py-3PEBTD COF (Figure 7.4q) [88]. In addition, Zhang et al. examined another CT complex based on a 2D-COF with a distinctive

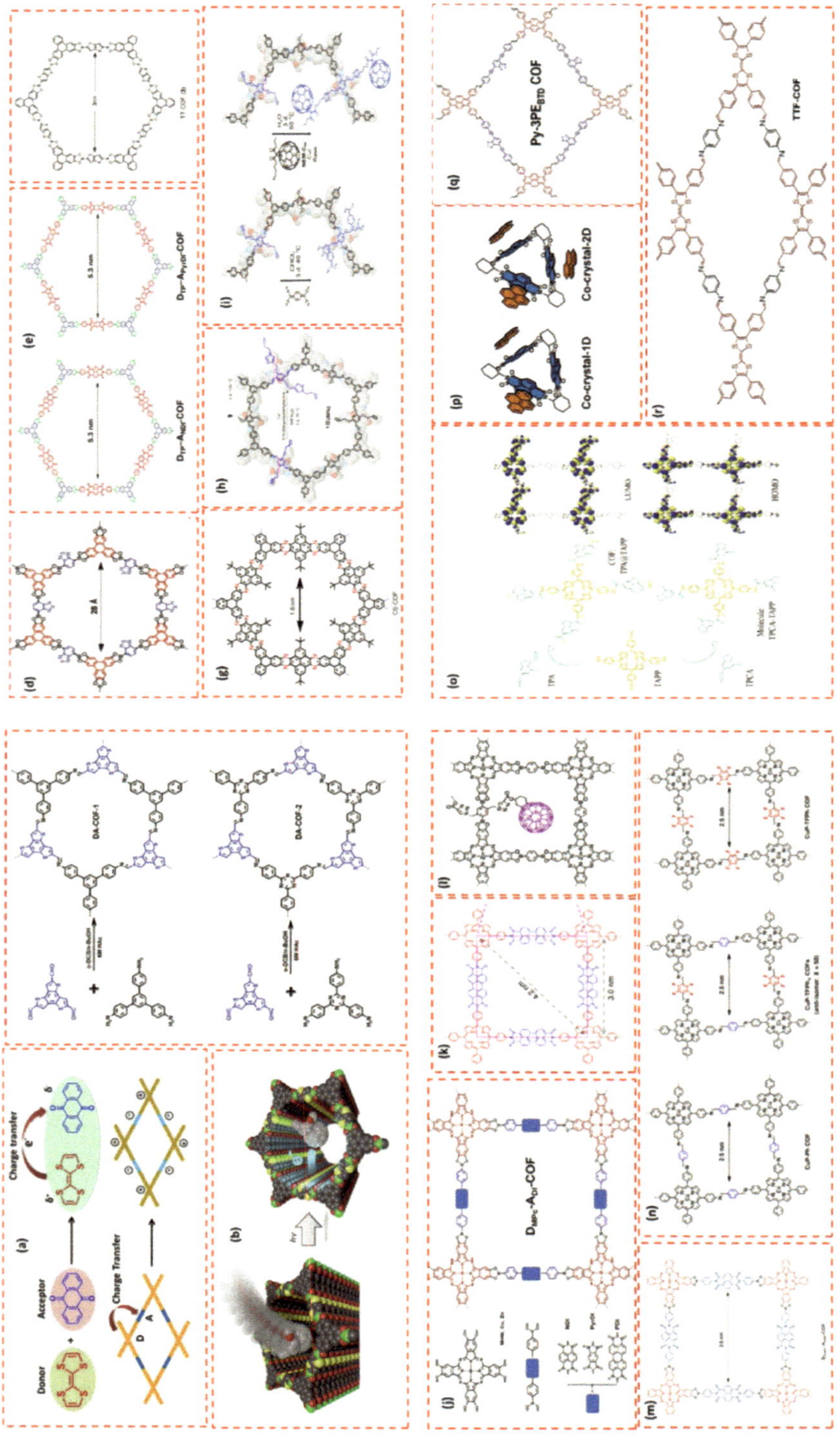

Figure 7.4 Skeletal structures of several COFs-based D-A (a) tetrathiafulvalene and pbenzoquinone compounds (Reprinted from [75] with the permission of RSC, license number: 1419845-1). (b) TT-COF and a PCBM (C grey, O red, B green, and S yellow) (Reprinted from [76] with the

permission of Wiley). (c) BTT for the electron– D and TPT/triphenyltriphenyl as A unit (Reprinted from [77] with the permission of Wiley). (d) triphenylene D units at the vertices and benzothiadiazole A (Reprinted from [78] with the permission of Wiley). (e) triphenylene as electron-D at vertices and naphthalene diimide or pyrromellitic diimide as A on edges (Reprinted from [79] with the permission of RSC). (f) TT-COF as D network to the encapsulated PCBM as A (Reprinted from [76] with the permission of Wiley). (g) CS-COF with C_{60} inside the open 1D channels (Obtained from open access paper) [80]. (h) 1,3-cycloaddition to form the D-A alignments using BPTA, DMTA, and TAPB (Reprinted from [81] with the permission of Wiley). (i) 3D buckyballs as the A component in a host matrix COF (Reprinted from [82] with the permission of Wiley). (j) DMPc-ADI COFs (Obtained from ACS publications) [83]. (k) PDI and free-base porphyrin (Obtained from open access paper) [84]. (l) D-A ZnPc-COFs with C60 (purple) (Obtained from ACS publications) [85]. (m) zinc phthalocyanine as D and naphthalene diimide as A (Reprinted from [86] with the permission of Wiley). (n) 2,3,5,6-TFTA and TA with CuP (TFTA) (Obtained from ACS publications) [87]. (o) TAPP (D) and TPA (A) (Reprinted from [27] with the permission of RSC). (p) Tri-NDI and Py as the A and D, respectively (Reprinted from [1d] with the permission of Springer). (q) Imine-linked rhombic lattice Py-3PEBTD COF (Obtained from open access paper) [88]. (r) doping TTF-COF with TCNQ (Reprinted from [89] with the permission of Wiley).

platform for the molecular design of electrical and optoelectronic materials. Tetrathiafulvalene-tetrabenzaldehyde (TTF-Ph-CHO) and p-phenylenediamine were combined to create TTF-COF through a condensation reaction. In this investigation, electron-donating TTF units and their derivatives, as well as TTF-COF, were disposed to doping with electron-A (TCNQ) (Figure 7.4r) [89].

Here, some recent researches on CT complexes through COFs are studied with more details summarized in Table 7.2. Industrial applications have not been reported for COFs due to the absence of charge separation efficiency. The significant strategy to improve the efficiency of charge separation of COFs is D-A alignment where charges can be separated through ICT, due to the formation of a polarized electric field [77]. In a recent study, a research group prepared two novel 2D-COFs constructed with a (diarylamino) benzene linker, resulting in the formation of a Kagome (kgm) lattice. The findings demonstrated the remarkable capability of these COFs in facilitating CT and diffusion processes. This can be attributed to their strong absorption of visible light, high degree of crystallinity, and large surface areas, reaching up to 1165 $m^2\ g^{-1}$. Strong reduction characteristics were provided by the diarylamine, which qualified these COFs to effectively reduce oxygen to create

H_2O_2. Also, metal-free materials have shown efficient photocatalytic solar transformations [90]. In addition, Gu et al. prepared an electron D-A COF (TTF-COF) by imine condensation of electron-A DAQ and electron-D TTF-fo as the building blocks. The research focused on creating planar building blocks that were specifically made for the synthesis of 2D electron D-A COFs. The focus was on analyzing the special characteristics of these COFs in assisting CT procedures. TTF-COF showed good thermal stability up to 300 °C and a high surface area of 729 m^2 g^{-1} [75]. Based on recent studies, perovskite solar cells constructed using DA-COFs have achieved an impressive maximum power-conversion efficiency of 23.19%, coupled with excellent humidity stability. In addition, Liu et al. presented research to explore methods for further enhancing the efficiency and stability of perovskite solar cells by incorporating D-A COFs. The incorporation of two D-A COFs into the FAPbI3 layer of perovskite solar cells resulted in improved porosity, high crystallinity, and outstanding stability. This integration effectively reduced the concentration of defects and minimized the depth of defect states within the perovskite layer. The results obtained from DA-COF-1 and DA-COF-2 showed a notable surface area of 1269 and 1561 m^2 g^{-1}, respectively, and thermal stability of 300 °C [77]. The extensive alignment of CT dipoles in two dimensions within the COF structure seems promising for potential applications in the fields of electronics and optics. Zhao et al. produced a D-A complex based on the Tri-NDI and the Py as the electron-D and A, respectively. The D-A complex was prepared via the self-assembly route using D-A interactions based on π-π bonds. The single crystal of the 2D-COF exhibited a more distinct CT effect compared to the 1D cocrystal [1d]. Also, understanding the emission mechanism in electrochemiluminescence (ECL) systems has been studied using a technique of control of CT between radical anions and cations. The D-A COFs not only offer a promising strategy for designing efficient ECL emitters but also contribute to advancing the understanding of ECL signal transduction in optoelectronic devices. Lei et al. designed a highly effective ECL emitter with tunable interarticular CT (IRCT) based on reticular structure D-A COF using triazine and triphenylamine units. The results showed the enhancement of 123 folds in ECL intensity for D-A COF rather than benzene-based COF. Moreover, the high crystallinity of COF structures and the influence of protonation on ECL behavior provided evidence for the dependence of ECL IRCT between D and A units. This correlation was further supported by DFT calculations. The ECL patterns observed in COFs exhibited dual peaks, which were achieved through a competitive oxidation mechanism mediated by IRCT. This mechanism involves coreactant-mediated oxidation at lower potentials and direct oxidation at higher potentials [91]. In addition, Luo et al. investigated a study on the novel application of COFs as a support and electron-D for the fabrication of CdS-COF nanocomposites. The research focused on investigating the potential use of these nanocomposites in solar energy transformations. CdS nanoparticles were deposited onto a covalent triazine-based framework (CTF)-1, utilizing a photo-deposition approach. The results demonstrated the effectiveness of CTF as both a photocatalyst and an electron-D. Remarkably, CdS-CTF exhibited superior hydrogen (H_2) production from water compared to the CdS/CTF nanocomposite, where CdS deposition was achieved through a solvothermal method after the H_2 evolution reaction under visible light irradiation. The improved activity of CdS-CTF was attributed to the presence of a larger number of active sites, which was a result of the confinement of small-sized CdS nanoparticles. Additionally, the better performance of CdS-CTF was ascribed to its rapid electron transfer rate (KET) and high injection efficiency. The KET amount of 0.18×10^9 S^{-1} for CdS-CTF was obtained which was 3.83 times faster than that of the CdS/CTF-1 nanocomposite. Similarly, the injection efficiency (η_{inj}) of CdS-CTF was recorded at 39.38%, which was 4.84 times higher compared to the CdS/CTF-1 nanocomposite [92]. Relatively fewer studies have focused on the direct integration of D-A molecules into core-linker arrangements within 2D materials. Most research has primarily involved the development of photoelectrically responsive COFs by incorporating either electron-D or A molecules into the COF pores, resulting in the formation of molecular CT pairs. In a related study, Crommie et al. investigated a coating of porphyrin-based molecular heterojunctions of COFs onto an Au substrate [93]. Furthermore, ensuring scalable thin film transfer, processing, and achieving wafer scale uniformity has remained a significant challenge [94]. Additionally, the accurate modification and rational design of D-A components for the integration of energy transfer (IET) and CT synergistic systems into 2D-COFs have been relatively less reported [27].

Table 7.2 Recent studies based on COFs for CT and D-A

CT-COF	Surface area m^2g^{-1}	Thermal stability (°C)	Bandgap energy (eV)	CT performance	Ref.
TAPD-(OMe)$_2$ COFs	1165	----	2.63	H_2O_2 production rate= 91.00 μmol h^{-1} g_{cat}^{-1}	[90]
TTF-COF	729	300	----	CT, capacitance = 752 F g^{-1} energy density = 57 Wh kg^{-1} power density = 858 W kg^{-1}	[75]
DA-COF-1	1269	300	----	Power conversion efficiency of perovskite solar cells = 23.19%	[77]
DA-COF-2	1561	300	----		[77]
Tri-NDI and Py	-----	-----	2.71	Potential applications in magnetism, ferroelectricity, electronics, and optical nonlinearity	[1d]
2D D-A COF	-----	-----	----	Suggesting a mechanism of phonon-assisted CT	[84]
CdS-COF	-----	-----	----	KET = 0.18 × 10^9 s^{-1} η_{inj} = 39.38% between CdS and CTF layers	[92]
t-COF	1372	------	1.06	ECL intensity increased by 123 times more than its benzene-based COF with minimal D-A contrast	[91]
b-COF	1689	------	----		[91]
a-COF	1364	-----	----		[91]

CT-COF	Surface area m^2g^{-1}	Thermal stability (°C)	Bandgap energy (eV)	CT performance	Ref.
COF-TPA@TAPP	6.034	-----	----	Efficient OPV film	[27]
C_{60}-metalloporphyrin COF	393	-----	1.81	TCNQ stacking within COFs showed a 32-fold increase in electron transfer rates than COF	[82]
TCNQ-covalently-linked COF	812	------	2.04		[82]
COF-420	----	----	1.92	Optoelectronic properties	[93]

7.3.3 CT Based on MOFs

MOFs belong to a specific category of crystalline porous materials that are characterized by the interconnection of metal ions (or nodes) through multitopic organic linkers. MOFs are recognized for their remarkable diversity, offering a wide range of materials with exceptionally high porosities and internal surface areas. Currently, there are approximately 20,000 known compounds within the MOF family [4a, 95]. Therefore, the advantages and disadvantages of CT based on MOFs can be described as follows. MOFs exhibit precise and well-defined structures, allowing for a controlled arrangement of components. This structural precision enhances electronic properties and facilitates efficient charge transport. MOFs offer tunable properties through the modification of their constituent metal ions and organic ligands. This tunability allows researchers to tailor the material's properties for specific CT requirements. Many MOFs possess high surface areas due to their porous structures, providing a larger interface for CT. This characteristic enhances overall conductivity and is beneficial for applications such as energy storage. The design of ligands in MOFs can be versatile, enabling the incorporation of various functional groups. This versatility allows for the enhancement of charge transport properties. MOFs generally exhibit good chemical stability, making them suitable for applications where exposure to different environments or harsh conditions is expected. However, some MOFs can be challenging to synthesize, and the optimization of synthesis conditions is crucial to achieving the desired properties [96, 96b].

Redox-active MOFs involve the transfer of electrons from one site to another coupled with the diffusion of counterions to maintain electroneutrality. In this area, the apparent diffusion coefficient (Dapp) is used to describe the overall charge transport for the combination of electron and ion diffusion [35c, 97]. MOFs can be tailored to exhibit semiconducting properties by incorporating alternating electron-D and electron-A molecules, like the structure of crystalline organic solids. This adaptation allows for enhanced semiconducting performance in MOFs [98]. Mixed valency has been utilized to control the conductive properties of materials, specifically for CT mechanisms. By incorporating organic or metal moieties with distinct formal oxidation states, these compounds exhibit light absorption across the electromagnetic spectrum, displaying their vibrant and colorful nature. This behavior results from the increased oscillation of charge between different centers, resulting in rapid CT phenomena [58]. Marcus-Hush theory has emerged as a fundamental tool for analyzing these systems, finding widespread application. This theory offers valuable insights into both through-space (outer sphere) and through-bond (inner sphere) electron transfer processes, providing the basis for understanding electron transfer phenomena [99].

Figure 7.5 illustrates the skeletal configurations of various D-A complexes utilizing MOFs. Here, some D-A units are mentioned based on MOFs. Hupp et al. reported CT within the zirconium-based MOFs with tetratopic 1,3,6,8-tetrakis(p-benzoate)Py (TBAPy4−) linkers and Zr6(μ3-O)4(μ3-OH)4(H_2O)4(OH)4 nodes. They used TBAPy4− linkers as electron-D and linker-defined cavities as A. In the case of NU-901, the cavities illustrated a diamond shape and approximately 60% of them abled the hosting fullerenes (Figure 7.5a) [13a]. Schröder et al. successfully synthesized MFM-300(VIII) by employing [V2(OH)2(L)] as the starting material, where LH4 represented biphenyl-3,3,5,5-tetracarboxylic acid. The investigation of crystal structure showed a tetragonal system in which the VIII center was coordinated by six oxygen (O) donors. Four of these donors were derived from carboxylate groups, and the other two originated from bridging hydroxyl (OH) groups, known as μ2-OH (Figure 7.5b) [100]. Furthermore, Schatz et al. studied the ICT phenomena within MOFs featuring hexa-ZrIV nodes and tetratopic linkers derived from tetraphenyl benzene 1, tetraphenyl Py 2, and tetraphenyl porphyrin 3. They successfully synthesized NU-1000 MOF utilizing linkers derived from tetraphenyl benzene 1 and tetraphenyl porphyrin 3 (Figure 7.5c). Additionally, they synthesized NU-901 and NU-1000 MOFs by incorporating a tetraphenyl Py linker and the hexa-ZrIV node. NU-1000 showed hexagonal channels, while NU-901 exhibited triangular channels. These MOFs offered unique structural characteristics, such as hexagonal channels in NU-1000 and diamond-shaped channels in NU-901 (Figure 7.5d) [34c]. In another study, 2D layers of ([Fe(2,2′-bipyridine)$(CN)_4]_2$Co-(4,4′-bipyridine))·$4H_2O$ (Fe_2Co-MOF) were prepared using a wet chemical method by Long et al. The materials were constructed by Fe and Co centers, where the Fe center was coordinated by two nitrogen (N) atoms and four carbon (C) atoms from cyanide groups. Furthermore, the Co center was situated within an N6 octahedral unit. The connection between these centers was established through cyanide units, extending in the direction of the c-axis. Notably, each Co center was surrounded by four identical Fe centers (Figure 7.5e) [101]. The CT complex based on MOF-5 as one of the first CT frameworks was synthesized by Majima et al. using precipitating a mixture of $Zn(NO_3)_2$·$6H_2O$ and a solution of terephthalic acid. The photoinduced CT reactions were carried out to examine the performance of MOF-5_n. The investigation focused on the presence of TEA in acetonitrile and (methyl viologen, $MV^{2+)}$ in water as D and A, respectively (Figure 7.5f) [102]. Moreover, Morris et al. synthesized MOF-808, NU-1000, and NU-1003 using the solvothermal method. These three MOFs were then incorporated into films, doped with ferrocene (Fc), with varying pore sizes ranging from 15 to 47 Å, (Fc-MOF). To introduce Ferrocenecarboxylic acid (Fc-COOH) into the MOF powders, a widely utilized post-modification technique called solvent-assisted ligand incorporation (SALI) was employed. This well-established procedure showed the successful integration of Fc-COOH with the as-prepared MOF powders (Figure 7.5g) [35a]. In addition, Mahata et al. investigated the luminescence behavior of a cadmium based on the luminescent MOF (LMOF), denoted as $[Cd(C_{12}N_2H_8)(C_7N_1O_4H_3)]$ (1). The LMOF was synthesized using a solvothermal reaction procedure (Fig. 7.5h) [103]. In another study, D'Alessandro et al. synthesized two zinc (Zn(II)) frameworks, namely, $[Zn_2(BPPTzTz)_2(tdc)_2]n$ and

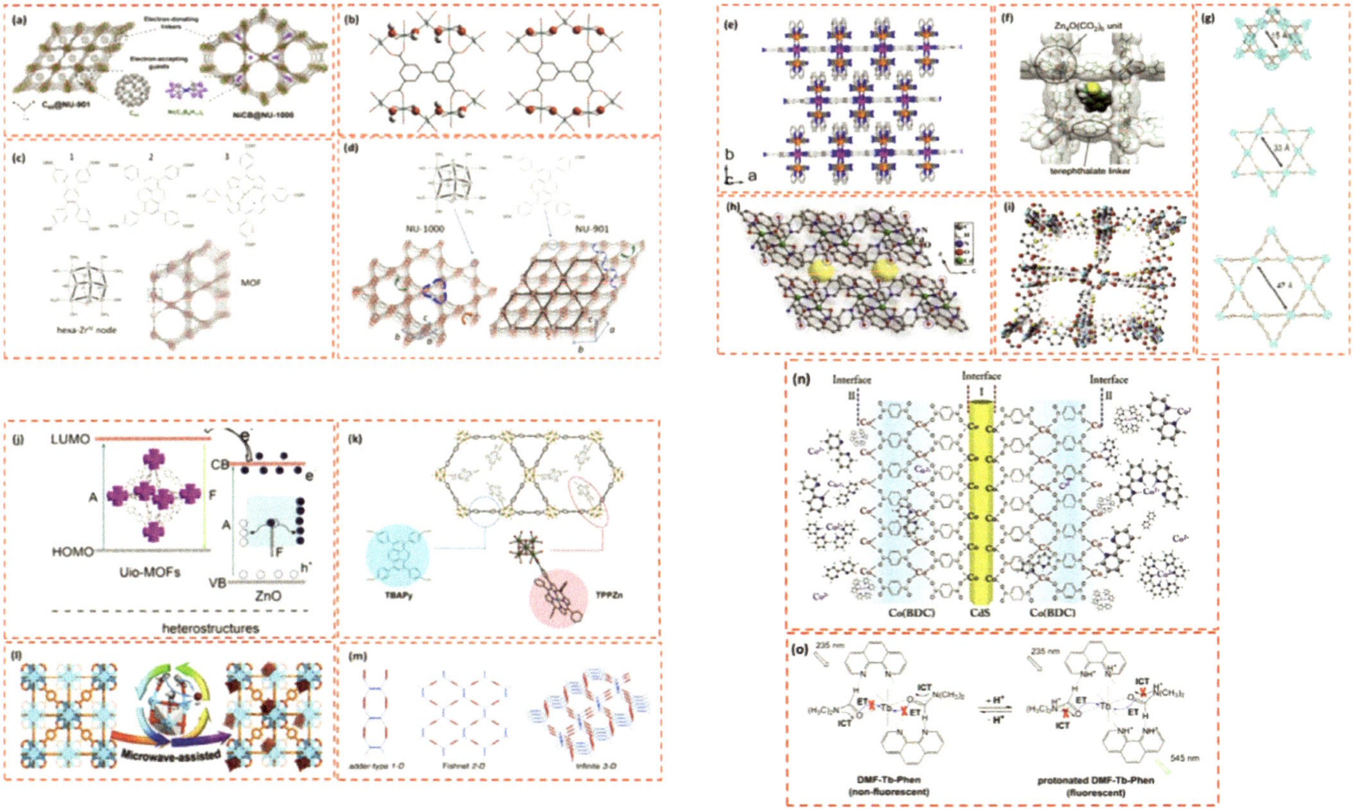

Figure 7.5 Skeletal structures of several MOFs-based D-A (a) TBAPy^{4-} linkers as D and cavities to host electron-accepting molecules (guests) (Obtained from ACS publications) [13a]. (b) MFM-300(VIII), hydroxyl groups are deprotonated to an O_2 bridge (red) (Obtained from

open access paper) [100]. (c) hexa-ZrIV node, and NU-1000 MOF containing linker 2 (Obtained from ACS publications) [34c]. (d) NU-901 and NU-1000 MOFs (Obtained from ACS publications) [34c]. (e) 2D layers of Fe2Co-MOF (Obtained from ACS publications) [101]. (f) MOF-5 (Obtained from ACS publications) [102]. (g) MOF-808, NU-1000, and NU-1003 (Obtained from ACS publications) [35a]. (h) $[Cd(C1_2N_2H_8)(C_7N_1O_4H_3)]$ (LMOF), yellow sphere= free space (Reprinted from [103] with the permission of RSC). (i) $[Zn2(BPPTzTz)_2(tdc)_2]n$(Obtained from ACS publications) [36]. (j) ZnO@Uio-MOFs heterostructures (Reprinted from [104] with the permission of RSC). (k) TPPZn@NU-1000 [105] (Obtained from ACS publications). (l) Fe-UiO-66 (Obtained from ACS publications) [106]. (m) [Ru2II,II] (red unit) and TCNQ (blue unit) (Obtained from ACS publications) [16]. (n) CdS@Co(BDC)-20 (Reprinted from [107] with the permission of Elsevier). (o) MOF based on DMF, Tb3+ and Phen (Reprinted from [108] with the permission of Elsevier).

$[Zn_4(BDPPTzTz)_2(tdc)_2]_n$. These structures were prepared through the heating of BDPPTzTz or BPPTzTz ligands with $Zn(NO_3)_2{\cdot}6H_2O$ and 2,5-thiophene dicarboxylic acid (tdc) (Figure 7.5i) [36]. In another study, ZnO@Uio-MOFs heterostructures were prepared by Yan et al. using doping semiconductor ZnO in Uio-MOFs. A novel fluorescence sensor based on a heterostructure was developed for the detection of volatile aldehyde gases (such as FA, AA, and ACA) at room temperature after the post-synthesis incorporation of Eu^{3+} (Figure 7.5j) [104]. In addition, Deria et al. presented a high-density D-A system by integrating a complementary pigment resulting from tetraphenylporphyrinato zinc(II) (TPPZn) into the 1D pores of a MOF called NU-1000, which was derived from 1,3,6,8-tetrakis(p-benzoic acid)Py (H_4TBAPy). The anchoring of the pigment within the pores occurred during the synthesis process (Figure 7.5k) [105]. Furthermore, Jiang et al. studied a well-known MOF called UiO-66 modified by introducing Fe^{3+} ions onto the Zr-oxo clusters, resulting in the formation of Fe-UiO-66. This modified MOF exhibited enhanced light absorption in the visible spectrum, thanks to a metal-to-cluster CT (MCCT) mechanism (Figure 7.5l) [106]. Moreover, Miyasaka described a D-A MOF system that was created through the combination of organic polycyano molecules (TCNQ/N,N'-dicyanoquinodiimine (DCNQI)) and carboxylate-bridged paddlewheel-type diruthenium (II, II) complexes ($[Ru_2^{II,II}]$). The investigation focused on analyzing the structural properties of the complex in 1D, 2D, and 3D arrangements (Figure 7.5m) [38]. Also, Peng et al. utilized a Co(BDC) MOF to link a heterogeneous CdS photosensitizer with a molecular co-catalyst $[Co(bpy)_3]^{2+}$ in order to enhance photocatalytic CO_2 reduction reactions (CO_2RR). They used the chemical complex $[Co(bpy)_3]^{2+}$ as a redox-targeting co-catalyst and covered the surface of the CdS nanorods with a layer of Co(BDC) MOF. Triethanolamine (TEOA) was employed as an electron-D for the efficient performance of photocatalytic CO_2 reduction (Figure 7.5n) [107]. In another study, Chen et al. employed lanthanide small molecular ligands and Tb^{3+} ions to create a pH-sensitive MOF nanoparticle operated based on the intramolecular CT effect. The sensitive material utilized included Tb^{3+} ions, 1,10-phenanthroline (Phen), and DMF. DMF was selected due to its unique combination of electron-D and A units, which made it particularly favorable for generating ICT phenomena (Figure 7.5o) [108].

Here, some recent researches on CT complexes through MOFs are studied with more details summarized in Table 7.3. MOF structure with sub-molar concentration pigment assembly within 10–30 Å pores offered exceptional photon absorptivity, as well as distinctive exciton interaction and dynamics. These features were different from those generally seen in π-stacked organic crystals and aggregates [105]. Furthermore, Hupp et al. studied water-compatible MOFs and maximizing porosity using a related approach. They used the specified cavities of the $TBAPy^{-4}$ linkers to host electron-A molecules (guests) and used the linkers as electron-D. The host fullerenes were loaded into the cavities of NU-901. The original porosity of MOF was preserved and electronic conductivity changed from less than 10^{-14} to 10^{-3} S cm^{-1} [13a]. In another study, Schröder et al. focused on investigating the binding behavior of I_2 (iodine) in two redox-active MOFs: MFM-300(VIII) and its oxidized and deprotonated counterpart, MFM-300(VIV). The findings revealed the formation of I3– species within the MOF channels, along with approximately 30% oxidation of the VIII centers in the host framework. The adsorption of I2 in MFM-300(VIII) led to a host-to-guest CT, resulting in a significant increase in electrical conductivity ($\Delta\sigma$ = 700,000) compared to MFM-300(VIII). Also, I2@MFM-300(VIII/IV) showed a high thermal stability of 500 °C. However, no noticeable change in conductivity or host guest CT was observed with the adsorption of I_2 in MFM-300(VIV) [100]. In studies based on MOF templates, Li et al. investigated the synthesizing of Z-scheme Fe_2O_3@Ag-ZnO@C heterostructures (FAZ) from the Fe-MOFs. The experimental findings on the application of Fe_2O_3@ZnO nanointerface and carbon sheath as photocatalysts for the degradation of methylene blue

(MB) and tetracycline (TC) pollutants in wastewater demonstrated excellent stability and remarkable performance under visible light irradiation. The enhanced photocatalytic activity was attributed to efficient absorption and high transfer efficiency of photogenerated carriers. Additionally, the incorporation of a carbon sheath and the favorable biological compatibility of the system provided valuable insights for the development of high-performance Z-scheme photocatalysts based on MOFs for environmental treatment [109]. Moreover, 2D layers of Fe_2Co-MOF was achieved by Long et al. by employing the $[Fe(2,2'\text{-bipyridine})(CN)_4]_2Co\text{-}(4,4'\text{-bipyridine})\cdot 4H_2O$ precursor. The process involved the incorporation of single-chain magnetism and underwent an intermetallic CT phenomenon, leading to a transformation in the charge/spin configurations of Fe_2Co from Fe^{3+} LS-Co^{2+} HS-Fe^{3+} LS to Fe^{2+} LS-Co_3+ LS-Fe^{3+} LS (LS = low spin, HS = high spin) at approximately 220 K under ambient pressure. The experimental findings revealed that Fe_2Co-MOF exhibited ferroelectric behavior and single-chain magnetism at low temperatures, accompanied by magnetism, coherent phase transitions, ferroelectricity, and permittivity. This multifunctional material was controlled by pressure, exhibited diverse properties, and held potential for various applications [101]. The creation of a synthetic MOF-based artificial light-harvesting complex (LHC) system offered new possibilities, allowing separated charges to move along the 1D pore channel. This presented an opportunity to gain a better understanding of the underlying mechanisms and for future developments. Furthermore, Deria et al. investigated a high-density D-A system by incorporating a complementary pigment derived from TPPZn into the 1D pores of H_4TBAPy-derived NU-1000 MOF. In order to allow energy transfer (ET) from the excited state of the MOF (NU-1000*) to TPPZn and CT from the excited state of the porphyrin (TPPZn*), the D and A components were selected to have acceptable ground- and excited-state redox potentials. The experimental findings indicated that the emission of the MOF was effectively quenched at 460 nm, aligning with the anticipated Py-centered emission. Particularly, an emission peak centered at 670 nm from TPPZn was observed (with an energy transfer rate of roughly $4.7 \times 1011\ s^{-1}$) because of the energy transfer from the excited MOF to TPPZn. Additionally, the charge separation process (with a rate of $1.2 \times 1010\ s^{-1}$) was achieved through an artificial "special-pair" like system involving the excited TPPZn pigment and a nearby TBAPy linker [105]. Plus, Ying et al. fabricated Ag-decorated ultrathin Ni-MOF on Cu2O with studding on CT dynamics at the heterostructure. The results showed an NH3 yield rate of 4.63 μg h^{-1} cm^{-2} with a faradaic efficiency of 24.3% [132]. Furthermore, Peng et al. investigated a study on utilizing a Co(BDC) MOF to connect a heterogeneous CdS photosensitizer with a molecular co-catalyst $[Co(bpy)^3]^{2+}$ for enhancing photocatalytic CO_2 reduction reactions (CO_2RR). The study produced notable findings, including a CO yield rate of 22 μmol h^{-1}, a selectivity of 92.5%, and a quantum yield of approximately 0.7% at a wavelength of 420 nm. The improvement of the photocatalytic properties of CO_2RR resulted from highly efficient CT occurring at both the MOF/solution and CdS/MOF interfaces, as supported by kinetic and spectroscopic investigations. The efficient performance of CdS@Co(BDC) in CO_2RR was attributed to the highly efficient CT occurring at both interfaces. The heterogeneous photosensitizer and homogeneous co-catalyst were connected by the MOF layer, which additionally enriched the CO_2 reactant. This layer also acted as a reaction platform. The produced electrons moved easily over the CdS/MOF (surface area of 48 cm^2 g^{-1}) interface after excitation, where they concurrently absorbed CO_2 and the molecular co-catalyst. At the MOF/solution interface, the photoelectrons were physically absorbed by molecular co-catalysts and transported to the redox species. Through well-established mechanisms of ligand-metal CT (LMCT) and proton-coupled electron transfer (PCET), this process produced CO and H_2 as end products [107, 111]. Furthermore, Majima et al. investigated luminescence transitions and the ICT in MOF-5 nanoparticles (MOF-5_n) and the temperature dependency of the spectral characteristics. The experimental findings demonstrated a significantly higher efficiency of the oxidation reaction in MOF-5 compared to P-25 TiO_2 powder, which is widely used as a photocatalyst. This observation presented a valuable opportunity to gain insights into the contrasting photoinduced CT mechanisms between MOFs and semiconductors [102]. The modifiable structure and electronic properties of MOFs offered significant advantages for their controllable light harvesting, redox activity, and, thus, photocatalytic performance. Moreover, Jiang et al. presented innovative research on the enhancement of light harvesting and photocatalysis in MOFs by exploring an MCCT process. They investigated the potential of MCCT for improving the performance of MOFs as photocatalytic materials. The Fe-UiO-66 structure was produced using Fe^{3+} onto the Zr-oxo clusters for MCCT. The results showed visible-light-driven water oxidation based on its electronic structure and strong oxidizing power. The selective activation of challenging C-H bonds, such as toluene oxidation was achieved using the generated holes in Fe-UiO-66. These holes exhibited the ability to selectively convert H_2O into hydroxide radicals, enabling efficient oxidation reactions [106]. In addition, Miyasaka studied the design of D-A MOFs by the precise modulation of CT through the combination of carboxylate-bridged paddlewheel-type diruthenium(II, II) complexes ([$Ru^{2II, II}$]) as electron-D and polycyanoorganic such as TCNQ and DCNQI as A. The CT in this system, even in the covalently bonded pattern, was dependent on the ID and EA of the D and A units, respectively. Also, chemical modification of the D and A units was utilized to control these properties systematically [38].

Table 7.3 Recent studies based on MOFs for CT and D-A

CT-MOF	Surface area m^2g^{-1}	Thermal stability (°C)	Bandgap energy (eV)	CT performance	Ref.
C_{60}@NU-901	-----	-----	----	Electronic conductivity = 10^{-3} S cm^{-1}	[13a]
NiCB@NU-1000	-----	-----	----	Electronic conductivity = 2.7 × 10^{-7} S cm^{-1}	[13a]
I_2@MFM-300(VIII/IV)	-----	150	----	Electrical conductivity ($\Delta\sigma$ = 700000)	[100]
Cadmium-based LMOF	-----	470	----	Zn^{2+} causes an enormous luminescence turn-on	[103]
Fc-MOF-808	390	-----	----	$k_{e\text{-hop}} \approx 10^3$–$10^7$ s^{-1}; $k_{i\text{-hop}} \approx 10^{-3}$–$10^1$ s^{-1}	[35a]
Fc-NU-1000	1690	-----	----		[35a]
Fc-NU-1003	2032	-----	----		[35a]

(*Continued*)

Table 7.3 (*Continued*)

CT-MOF	Surface area m^2g^{-1}	Thermal stability (°C)	Bandgap energy (eV)	CT performance	Ref.
$[Zn_2(BPPTzTz)_2(tdc)_2]_n$	53.4	300	----	Anodic region, a broad process at ~1.0 V vs Fc/Fc^+	[36]
$[Zn_4(BDPPTzTz)_2(tdc)_2]_n$	-----	300	----		[36]
TPPZn@NU-1000	1540	100	NU-1000 = 3.10 TPPZn = 2.17	k_{EnT} = 4.7× 10^{11} s^{-1} k_{CT} = 1.2 × 10^{10} s^{-1}	[105]
Ag/Ni-MOF/Cu_2O	-----	-----	----	NH_3 yield rate = 4.63 μg h^{-1} cm^{-2} Faradaic efficiency =24.3%	[110]
CdS/MOF: CdS@ Co(BDC)-20	48 cm^2 g^{-1}	-----	Cds = 2.35	Photocatalytic CO_2 reduction reactions (CO_2RR), CO yield rate = 22 μmol h^{-1} Selectivity = 92.5%	[107]
TiO_2/NH_2–UiO-66	394	900	----	CO gen = 0.44 ± 0.09 μmol $g_{cat}^{-1}h^{-1}$ CO_2 ad = 0.53 (mmol g^{-1})	[112]
Fe-UiO-66	-----	-----	----	Photocatalytic oxidation of hydrocarbons	[106]
Cu–ETTC MOF	983.6	-----	----	The intramolecular LMCT acts as a fluorescent switch in MOF nanomaterials	[113]

7.4 Synthetic Building Frameworks

7.4.1 CMPs Synthesis Methods

The 3D structure of semiconducting polymers in CMPs consists of rigid aromatic groups interconnected by single, double, or triple bonds [114]. Since the introduction of D-A complexes based on CMPs, there have been numerous attempts at supramolecular chemistry. These include the development of supramolecular photosystems [115], the synthesis of rotaxanes and catenanes [116], the fabrication of synthetic ion channels [117], the exploration of liquid crystals [118], and numerous other innovative strategies [119]. The synthesis of CMPs primarily depends on the solution phase polymerizations, employing various techniques such as the Suzuki reaction [120], Sonogashira-Hagihara reaction [121], oxidative coupling reaction [69, 122], Yamamoto reaction [50, 120a, 123], cyclotrimerization reaction, Schiff base reaction [124], phenazine ring fusion reaction [53], and Friedel–Crafts reaction [125]. CMP manufacturing is frequently challenging because most CMPs are not processable and insoluble in most solvents [46]. The creation of solution-processable CMPs to enable the manufacture of thin films would represent a substantial advancement in electrical devices. This presents a considerable challenge for synthetic chemistry due to the poor inherent solubility of the rigid and cross-linked structure, which limits the formation of high-quality CMP films [126]. Besides, CMPs are produced using solution methods, based on the well-established Suzuki reaction [127], Yamamoto reaction [128], Sonogashira-Hagihara reaction [129], the oxidative coupling reaction, etc. [122a]. In addition, the typical techniques such as hyperbranched soluble CMPs [130], layer-by-layer cross-linking [131], and surface-initiated synthesis [132], were used to fabricate CMP films. The challenges in these processes limit the applications of CMPs in organic electronics. A study in this area emphasized that achieving processability and device integration was a challenge [133]. Then, the results obtained from the literature revealed methods such as Yamamoto couplings [134], Sonogashira–Hagihara [135], and oxidative polymerizations [136]. The limitation of solubility for CMPs resulted in low reactivity and, consequently, the formation of structural defects in the products. The research demonstrated that electropolymerization is an effective method for fabricating CMP films, as it allows for the direct formation of films on electrodes without the need for polymer powder. With this method, thin films can be produced quickly and in a short amount of time. Moreover, the electropolymerization approach avoids the need for catalysts, high temperatures, and an inert atmosphere typically required in other methods of synthesizing bulk CMPs [70].

7.4.2 COFs Synthesis Methods

These highly ordered and porous polymers are synthesized through the polycondensation of geometrically designed building blocks. Alternatively, by incorporating dynamic covalent bonds and symmetry monomers, such as C2; aromatic diamine, dialdehyde, diboronic ester, diboronic acid pinacol ester, and so on, C4; porphyrin and phthalocyanine, COFs with diverse pore channels can be produced, offering tunable shapes and sizes. This integration of organic building blocks enables the creation of COFs with customizable pore architectures [137]. The extended frameworks of crystalline COFs are constructed by bonding monomers through proper linkages. In COF chemistry, several linkages have been used, such as boroxine [138], boronic-ester [78, 139], borosilicate [140], triazine [141], imine [142], hydrazone [143], borazine [144], phenazine [80], imide [145], amide [146], C=C [147], 1,4-dioxin [148], and hypercoordinate silicon [149]. The polycondensation between the D and A monomers in COFs should be employed to produce D-A COFs. So far, only a few boronic-ester [78, 79, 83, 85, 86, 150], imine [81, 82, 88, 151], and C=C-linked [152] to prepare D-A COFs have been described. COFs are designed using a range of building blocks, allowing for accurate tuning of their responsivity to visible light and bandgap characteristics [151i, 153]. Moreover, the well-structured framework and robust π–π interactions within COF layers enhance the process of charge separation and facilitate efficient charge transport [154]. The common synthetic methods to produce COFs are solvothermal [155], multicomponent reaction, mechanochemical synthesis [156], microwave synthesis [157], multistep synthesis, and linker exchange strategies [10b]. The crystallinity of COFs is greatly influenced by various factors throughout the synthetic procedure. Several factors affect the product, including the pressure maintained within the sealed vessel, the reaction temperature, the reaction time, the ratio of solvent mixtures employed, and the quantity of catalyst used [10b]. A considerable number of imines-linked COFs have been successfully synthesized with the aid of aqueous acetic acid (AcOH) as a catalyst. However, in some instances, the acidic conditions posed compatibility issues with specific components [10b, 158]. Furthermore, protocols utilizing self-correction and linker-exchange mechanisms have been created, relying on the reversible nature of covalent bonds in thermodynamic equilibrium. These protocols are designed to achieve structures that possess both crystallinity and stability [159]. According to existing literature, traditional methods for synthesizing boronic ester or imine-linked COFs provide difficulties related to the multivariant free reactive groups such as catechol, boronic acid, aldehyde, and amine. Some of these challenges are susceptibility to oxidation, limited accessibility, and low solubility, which create obstacles to the development of COFs for various applications. Achieving accurate stoichiometry of the reactive sites provides several benefits, including improved solvent adaptability, enhanced reproducibility, and crystallinity of COFs [23]. A recent advancement in COF synthesis is the two-in-one strategy, which allows for the straightforward preparation of highly crystalline 2D imine-linked COFs with excellent reproducibility and solvent adaptability. The main idea behind this strategy is to combine two different functional groups (reactive sites) into a single molecule. In conventional imine-linked COF synthesis, formyl and amino groups play a vital role but are typically presented in separate monomers [160]. Traditionally, COFs have been primarily obtained as insoluble and unprocessable solids, posing important challenges in their integration into devices. However, the 3D printing technology has provided a useful post-processing solution for COF powders, enabling the fabrication of tailored multicomponent COF monoliths with desired shapes and structures. This approach has created new possibilities for the practical utilization of COFs in various applications [161]. Moreover, there are even more efficient techniques available for processing COF powders into films. These include the bottom-up strategy implemented through solvothermal methods [162], as well as the top-down approaches involving either physical or chemical exfoliation [163]. These techniques have successfully enabled the use of COF films in several applications.

7.4.3 MOFs Synthesis Methods

Traditionally, MOFs have been prepared through hydrothermal [164], and solvothermal [165] synthesis routines, which often require several hours to days of reaction time. However, alternative synthesis methods have been explored to reduce the synthesis time and obtain smaller and more uniform crystals. These methods include microwave-assisted [166], sonochemical [167], electrochemical [168], chemical precipitation [169], and mechanochemical [170] approaches. Some of the common synthesis methods exist for the production of porous frameworks as mentioned above for COFs. Recently, the effectiveness of microwave irradiation in synthesizing organic polymers and MOFs has been widely recognized. In addition, mechanochemical synthesis is a straightforward and cost-effective approach that has been successfully utilized in the synthesis of MOFs. Furthermore, it has been proven to be suitable for the synthesis of COFs as well [156]. MOFs can be synthesized in various forms such as pellets, isolated single crystals, polycrystalline powders, mixed-matrix membranes, thin films, and more. The choice of the specific form depends on the desired morphology and intended application of the MOFs. In addition, MOFs have been synthesized using various design strategies, including the infiltration of guest molecules or nanocrystals, as well as the formation of 1D chains composed of alternating metal sulfur or metal-oxygen atoms [171], π-stacking in aromatic linkers [34a], and penetration with nanocrystals or guest molecules [172].

7.5 Functionalization of CMPs, MOFs, and COFs for Semiconductors

In order to address the limitations, the introduction of functional moieties into the structure of porous frameworks (CMPs, COFs, and MOFs) to enhance both CT performance and redox activity has been studied. Functionalization of CMPs, COFs, and MOFs involves modifying the structures of these porous materials to enhance CT performance can involve:

(i) Backbone modification: It helps in modifying the polymer backbone structure influences on the semiconducting properties and adding conjugated moieties or specific functional groups that can enhance charge transport and optoelectronic characteristics.

(ii) Porosity control: It tunes the porosity and surface area can optimize their performance in semiconductor applications and controls the pore size and distribution that can facilitate the efficient diffusion of charge carriers and enhance device performance.

(iii) Doping and guest molecule incorporation: It helps in doping with specific atoms or incorporating guest molecules can modify their electronic properties and enable tailored semiconductor behavior. By incorporating semiconductor molecules or nanoparticles as guest species, they exhibit semiconductor behavior. Also, the band structure, carrier concentration, and conductivity of the MOF can be changed by incorporating metal ions or clusters of different valences. In addition, MOFs can encapsulate guest molecules within their pores, which modify their properties.

(iv) Post-synthetic modification (PSM): The porous frameworks can be functionalized through post-synthetic reactions to introduce specific functional groups onto their surface or within their pores. This can enable the fine-tuning of electronic properties, such as bandgap and charge transport behavior, for semiconductor applications. Also, the addition of conjugated π-systems through PSM enhances the electronic properties of MOF.

(v) Ligand engineering: The tuning of electronic properties in MOFs can be achieved through the design and synthesis of organic ligands containing specific functional groups. Ligands with electron-donating groups influence charge transport behavior, bandgap, and conductivity.

Here, the examples of functionalization of CT complex through the synthetic porous frameworks are presented. EL-Mahdy et al. synthesized a range of semiconducting D-A CMPs with varying levels of sulfur and nitrogen incorporation and different degrees of conjugation. These CMPs were specifically designed to facilitate efficient photocatalytic water splitting, leading to the generation of H_2 and O_2. The Cz units were utilized in CMP synthesis for photocatalysis, owing to their exceptional properties such as electron donation, charge transport, and ease of post-functionalization [58]. Researchers have also investigated additional structural modifications of conjugated porous polymers (CPPs) to improve their wettability, consequently enhancing their photocatalytic capabilities. Specifically, CPPs based on truxene-fluorene with amino side chains (Tr-F3N) and lateral octyl (Tr-F8) were tailored for the purpose of photocatalytic water splitting. By introducing hydrophilic amino side chains into Tr-F3N, the wettability of the material was notably enhanced, leading to a higher HER of 500 μmol g^{-1} h^{-1}. In contrast,

Tr-F8, functionalized with hydrophobic octyl groups, exhibited lower HER. These results emphasize the significance of amino groups in improving photocatalytic performance by promoting water absorption at the molecular level, accelerating the HER process [173]. COFs can be molecularly functionalized to improve CT properties by altering the initial linker. It has been demonstrated that adding functionality like triazine, benzothiadiazole, and porphyrin enhances CT kinetics and encourages effective charge carrier separation in COFs. These molecular-level alterations provide a way to enhance the CT efficiency of COFs and broaden the range of possible applications [153a, 174]. The enhanced photocatalytic performance of COFs was achieved through the incorporation of alternating D-A moieties into extended frameworks [150c, 173g]. The molecular-level modification of the structures has demonstrated significant potential in optimizing COFs for semiconductor applications [90]. Furthermore, Banerjee et al. described the incorporation of functional moieties exhibiting CT characteristics as a successful approach to improving the electrochemical performance of COFs. This enhancement was attributed to the subsequent increase in electron mobility. The outcomes showed the electron D-A units were an impressive way to improve CT [175]. In addition, Bein et al. prepared a COF using two linear monomers: dialdehyde benzo[1,2-b:4,5-b']-dithiophene-2,6-dicarboxaldehyde (BDT) and a tetraphenylethylene derivative with four amine functional groups (1,1,2,2'-tetra-p-aminophenylethylene, or ETTA. These building blocks were used to synthesize COF structures by assembling conjugated polyimide sheets through aromatic amine functionalized tetraphenylethylene and thiophene-based dialdehyde units. These sheets were arranged in a 3D stack to create photoactive porous frameworks. The researchers initially examined the optoelectronic properties of BDT-ETTA COF films and found that they exhibited suitable bandgap energy, as well as appropriate positions for the HOMO and lowest LUMO. These properties indicated that the COF films were utilized for HER applications [176]. Hupp et al. explored the potential of postsynthetically functionalized MOFs to exhibit 3D arrangements of densely packed active sites that were effectively separated [13a]. Deria et al. reported a PSM study on MOFs, specifically focusing on anchoring complementary pigments formed from TPPZn within the 1D pores of the H_4TBAPy-derived

NU-1000 MOF. This process resulted in the formation of a densely packed D-A system due to the binding capacity of the octahedral Zr_6-oxo cluster, which accommodated up to twelve carboxylates. Each node of 8-connected Zr-MOFs was successfully decorated with four carboxylate-terminated pigments or redox counterparts. The SALI technique was employed for the PSM. The process of SALI included the substitution of a pair of terminal hydroxyl and aqua ligands with incoming carboxylate ligands. As a result, a functionalization occurred in a mesoporous ZrIV-oxo MOF, leading to the formation of a 1D pore channel. An artificial LHC-like system can be created using this modified MOF and a complementing redox partner [105]. Furthermore, Mahjoub et al. reported a novel photocatalyst called amine-functionalized MOF@Sm_2O_3–ZnO (abbreviated as MOF@ Sm_2O_3–ZnO NCP) that was incorporated into MOFs. This NCP demonstrated exceptional performance as an active and stable photocatalyst under visible light, effectively degrading AMX (an unspecified substance). This study represented the first exploration of the photocatalytic efficiency of NH_2–MOF@Sm_2O_3–ZnO NCPs in the degradation of antibiotics [177]. Functionalization of CMPs, COFs, and MOFs for semiconductor applications is an active area of research. Continual efforts focus on optimizing the design and functionalization strategies to enhance their performance and expand their application range in various electronic devices, such as sensors, transistors, photovoltaics, and energy storage systems.

7.6 Theoretical Studies of CT Complex

The theoretical studies focus on CT processes and rationalize the complex photophysical by DFT calculations. Here, some recent researches in the CT through CMPs, COFs, and MOFs based on DFT calculations are described. DFT, a fundamental quantum computational theory, offers a quantum mechanical approach to investigating the electronic structure of systems with multiple electrons. In recent years, it has become increasingly important in directing the functional design of CMP materials [16, 178]. Zhang et al. used the density functional tight-binding approach to analyze the electronic structure and adsorption characteristics of single-walled CMP nanotubes that incorporate substituted benzene and ethylene units, based on DFT investigations. According to computational studies, decreasing the junction agent or adding substituents (such as AOH or ANH_2 substituents) at the node effectively reduced the bandgap of the nanotubes. This modification led to a larger extended π-conjugation system, reinforcing the bonding strength between the adsorbent and adsorbates. Consequently, the energy-based adsorption capacity was enhanced, promoting improved adsorption performance [179]. In another research, Xu et al. utilized a spin-polarized DFT approach to design a new 2D-CMP composed of thiophene chains and porphyrin units. They added several metal atoms (Fe, Mg, Mn, and Cu) to the porphyrin ring center to increase its capacity to reduce CO_2 [195]. Additionally, the stability of the CMPs under various environmental conditions by investigating computational analysis was determined. The results revealed that the Fe-modified CMP (Fe-CMP) demonstrated the most outstanding catalytic activity for CO_2 reduction when using transition state (TS) search calculations. The evaluation of the energy band structure and the distribution of frontier orbitals provided further confirmation of the semi-metallic properties and high electron occupancy, which played a significant role in influencing photocatalytic performance. Furthermore, the study investigated the potential mechanism of CO_2 photocatalytic reduction. It revealed that the conversion of CO_2 into methane occurred through the COOH pathway and involved several thermodynamic steps [180]. Additionally, Maji et al. also used time-dependent density functional theory (TDDFT) calculations on reduced cluster models of $F_{0.0}$CMP and $F_{2.0}$CMP, which were labeled as 2D and 2D1A, respectively, to understand the experimental data. The 2D structure was composed of two TPE molecules and a phenyl spacer. On the contrary, 2D1A was constructed by combining two TPE units and one 9-fluorenone unit, with the A-portion being separated from each TPE unit by a phenyl spacer. The TDDFT results for 2D showed that the lowest energy excitation mainly exhibited πHOMO-π*LUMO features, with both frontier molecular orbitals (FMOs) dispersed along the whole length of the molecule [68]. Furthermore, Zhang et al. performed DFT calculations to investigate the CT interactions between the anthracyl and the Bodipy framework in CMPs. Gaussian software was used to prepare these calculations at the B3LYP/6-31G (d) energy level (Figure 7.6). The findings indicated that the absorption of CT from anthracyl

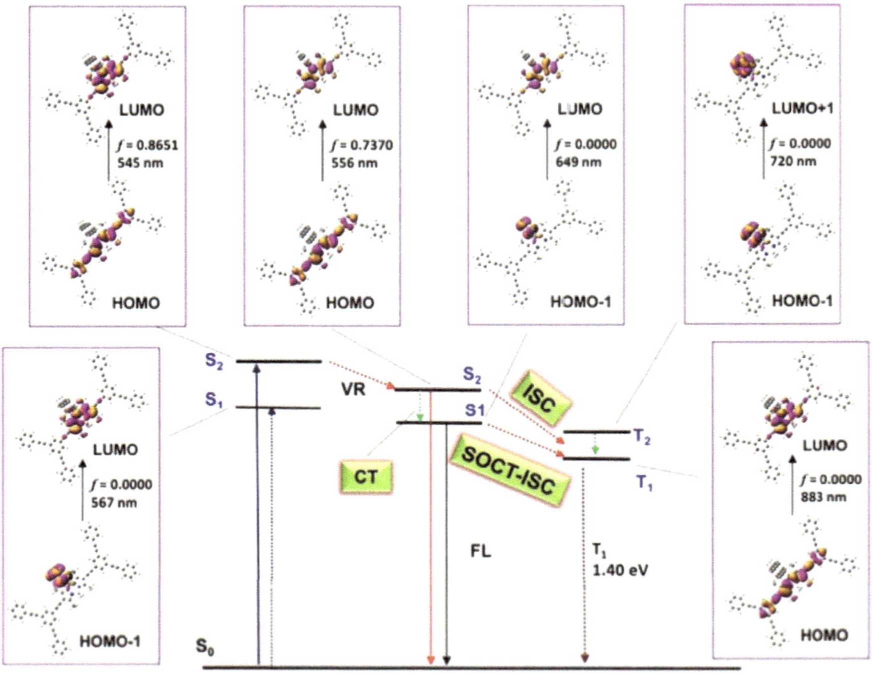

Figure 7.6 The schematic of frontier molecular orbitals engaged in the excitation, fluorescence (FL), and triplet excited state of the monomer in CMP-2, with VR representing vibrational relaxation. The left section displays the UV/Vis absorption, while the middle section exhibits the FL emission. The right section presents the triplet excited state (Reprinted from [62] with the permission of Wiley).

to the Bodipy framework, as well as to the Bodipy unit within the framework, resulted in two transitions from S_0 to S_1 state and S_0 to S_2 state. Furthermore, the oscillator strength (*f*) provided valuable information about the transition probability, with a value of 0.0000 for the S_0 to S_1 transition and 0.8651 for the S_0 to S_2 transition. This observation indicated that the Bodipy framework played a crucial role in absorbing visible light in CMP-2. The localized singlet state (S_2) of Bodipy was populated after vibrational relaxation, taking an electron from anthracyl to create the 1CT state. These findings showed that adding anthracyl to CMPs created a new route for improving electron-hole separation, driving the photoinduced electron transfer reaction. Moreover, the S_2 state showed the ability to directly undergo intersystem crossing (ISC) to the T_1 state, and this process was facilitated by polymerization and the appropriate energy level difference. The transition from S_2 to S_1 to T_1 introduced an additional ISC pathway through SOCT-ISC, which was applied as a suggestion to further enhance the efficiency of ISC. Additionally, the Bodipy framework within CMP-2 played a central role in both light absorption and catalysis due to the localization of the lowest triplet state on it. The presence of anthracyl in CMP-2 promoted CT between Bodipy and anthracyl, resulting in an improved ISC process and enhanced electron-hole separation [62].

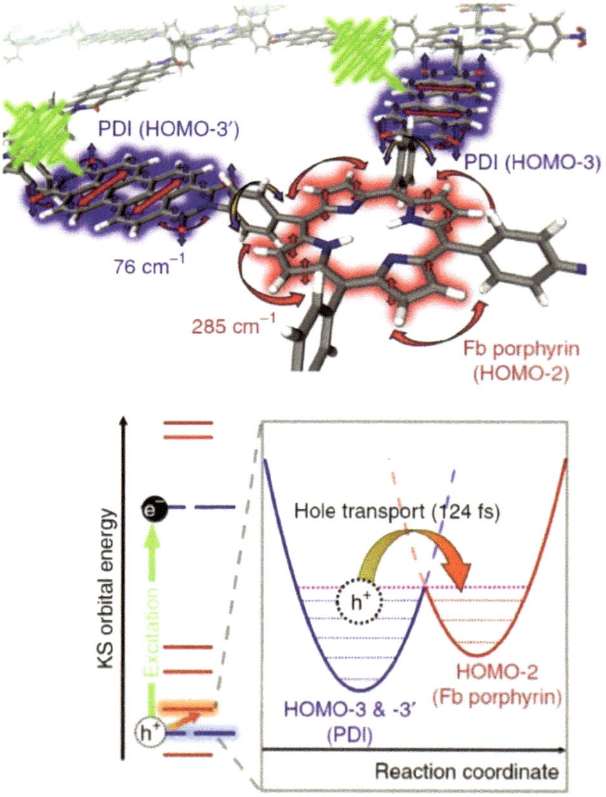

Figure 7.7 The charge carrier dynamics in photoinduced COFs (Obtained from open access paper [84]).

In a separate study, the HOMO and LUMO energy levels were determined using DFT calculations, and the bandgap values were found to agree with experimental measurements. El-Mahdy et al. performed DFT calculations at the B3LYP level, utilizing the 6-31G(d,p) basis set, to investigate the structure-property relationships of four CMPs: Cz-TPT, Cz-BT, Cz-TzTz, and Cz-Py. They focused on analyzing the energy gaps between the HOMO and LUMO. Theoretical calculations revealed that the TPT, BT, TzTz, and Py units as electron-A, and the Cz units as electron-D. The electron densities of the LUMO states in the CMPs were predominantly located on the TPT, BT, TzTz, and Py units. In contrast, the HOMOs were distributed across both the Cz units and the TPT, BT, TzTz, and Py segments. This conjugation between electron-D Cz units and electron-A TPT, BT, TzTz, and Py resulted in an effective delocalization of the HOMO levels. This extended delocalization has the potential to facilitate intramolecular CT processes, ensuring high mobility within the CMPs [58]. Furthermore, Ihee et al. used nonadiabatic molecular dynamics with femtosecond optical spectroscopy to study the dynamics of photogenerated charge carriers in 2D D-A COFs. This result revealed that the creation of polaron pairs occurred via rapid hole transfer within the layer, which was connected to the coherent oscillations of the 2D crystal structure. These findings proposed a mechanism that involved phonons assisting in CT (Figure 7.7) [84].

Moreover, based on DFT calculations for CT-COF, Kong's group proposed the nitrogen atoms within the triazine rings functioned as active sites for photocatalysis. They further suggested that the D-A structure enhanced the separation of excitons and facilitated the efficient transport of charge carriers. Through isotopic experiments using CO_2 and H_2, the formation of CO and O_2, respectively, was confirmed, indicating the photoreduction of CO_2 and the oxidation of H_2O [129g]. In addition, theoretical studies of CT based on MOF can refer to the report of Morris et al. They created films of ferrocene-doped MOFs (Fc-MOF) adsorbed on conductive substrates with different pore sizes ranging from 15 to 47 Å to study redox-active MOFs. They independently quantified the numerical results of electron and ion diffusion coefficients, as well as electron and ion transfer rate constants, by applying a theoretical model to the chronoamperometric responses of the three Fc-MOFs. The results were including Di ≈ 10^{-16}–10^{-12} $cm^2\ s^{-1}$ and De ≈ 10^{-12}–10^{-7} $cm^2\ s^{-1}$. An increase in ki-hop, a decrease in ke-hop, and an increase in the overall CT rate constant were related to the increasing pore size of the MOF [35a]. Moreover, D'Alessandro et al. investigated CT in MOFs and explored the phenomenon of through space intervalence CT (IVCT) as a novel mechanism for CT in MOFs. The findings indicated the formation of a mixed valence state through electrochemical or chemical reduction, attributed to the aromatic stacking interactions within Zn(II) frameworks that incorporated cofacial TzTz units. The ρ was quantified through the utilization of both ex situ and in situ spectro and electro-electrochemical methods. Computational modeling of the optical data, in combination with the application of Marcus-Hush theory to the through-space IVCT bands observed in the mixed-valence framework, facilitated the determination of the extent of CT. Experimental and computational studies identified a distance-dependent through space electron transfer phenomenon. The final results presented a novel perspective on electron transfer mechanisms in 3D coordination space [36]. Additionally, Schatz et al. studied computational investigations of ICT in NU-901 and NU-1000 employing tetratopic carboxylate linkers and hexa-ZrIV nodes formed from molecules with the names tetraphenyl benzene 1, tetraphenyl Py2, and tetraphenyl porphyrin 3. The main parameters (G^0, and t) were evaluated to determine the CT propensities in MOFs based on the chemical structures of the linkers. This qualitative analysis showed that t was enhanced and decreased by delocalizing the HOMO/LUMO on terminal substituents. On the other hand, weaker counterion binding decreased G^0 and raised the chance of CT [34c].

7.7 Electrical Conductivity of CMPs, COFs, and MOFs

Electrical conductivity results from the multiplication of the number of charge carriers (n) and their mobility (μ). This phenomenon is prominent in materials with partially filled energy bands and effective electronic coupling between neighboring building blocks [181]. In the realm of materials science, planar π-conjugated electron-D and A moieties have found widespread application in the development of p- and n-type semiconducting COFs. Most semiconducting COFs are rooted in D building blocks, some of which can form stable cation radicals through oxidation. Several of these conjugated building blocks have been extensively studied for their utility in organic electronics, including TTF [204], triphenylenes (TP) [183], porphyrins (Por) [73], pyrenes (Py) [184], hexabenzocoronenes [207], phthalocyanines (Pc) [186], as well as more recent additions such as Wurster-type tetraphenyl-p-phenylenediamine compound (W) [187] and triangulene (TANG) [188]. Moreover, electron-D building blocks including naphthalene diimides (NDI) [79], benzothiadiazole (BTD) [189], pyridines [190], and triazines [191] have also been harnessed in the construction of n-type or ambipolar semiconducting COFs. These advancements represent significant progress in the field of organic semiconductors, paving the way for diverse applications in electronic devices and beyond [181]. On the other hand, based on recent research by Dinca et al., four established strategies were employed to achieve high conductivities in materials. Firstly, the through-bond approach involves the creation of continuous chains of coordination bonds between metal centers and functional groups of ligands, establishing pathways for charge transport. Secondly, the extended conjugation approach entails the formation of large delocalized systems by integrating metals and entire ligands. Thirdly, the through-space approach capitalizes on π–π stacking interactions between organic moieties. Lastly, the guest-promoted approach leverages the innate porosity of MOFs and exploits host–guest interactions to enhance conductivity. These strategies represent key avenues in the pursuit of highly conductive materials, each offering unique pathways for optimizing electronic properties [192, 192b]. On the other hand, the pursuit of elevated conductivities in MOFs draws inspiration from well-established domains. Two primary methodologies, the through-bond and through-space approaches, play pivotal roles in this endeavor. The through-bond approach centers on enhancing the bonding between the metal and ligands, facilitating robust charge delocalization. Conversely, the through-space approach concentrates on noncovalent interactions among organic fragments, with a specific emphasis on π–π stacking, thereby establishing uninterrupted charge transport pathways [194c, 193]. It should be mentioned that the different electroactive building blocks have been classified according to the electronic nature and main functionalities of the ensuing frameworks, namely: (i) redox-active building blocks for tunable electrically-conductive frameworks and energy storage;

(ii) light-responsive building blocks for optoelectronics; and (iii) magnetically active organic radical building blocks [193]. Some research studies on the electrical conductivity of COFs and MOFs have been reported, which have been recorded in a wide range (10^{-10} to 10^{-1} S cm^{-1}). It results from prepared in different structures including composites and functionalized to improve electrical conductivity. In this area, a straightforward in situ solid state polymerization procedure was carried out inside the COFs nanochannels by Awaga et al. They successfully produced a nanocomposite by redox-active AQ-COF and conductive poly(3,4 ethylenedioxythiophene) (PEDOT). The PEDOT chains acted as efficient carriers for electron transport within the nanochannels, resulting in a PEDOT@AQ-COF nanocomposite. This nanocomposite exhibited outstanding electrical conductivity, reaching 1.1 S cm^{-1} with a surface area of 131 m^2 g^{-1}, and demonstrated significantly enhanced performance in faradaic energy storage. The combination of PEDOT and AQ-COF resulted in synergistic effects, enabling improved conductivity and superior energy storage capabilities in the nanocomposite. AQ-COF exhibited a conductivity of approximately 1×10^{-10} S cm^{-1}, indicating its insulating nature. In comparison, the conductivity of 1.1 S cm^{-1} and 1.9 S cm^{-1} were obtained for PEDOT@AQ-COF and PEDOT, respectively. According to the electrical conductivity, both PEDOT and PEDOT@AQ-COF were considered conductive materials. In particular, the conductivity of PEDOT@AQ-COF was approximately 10 times of magnitude higher than that of AQ-COF [194]. In addition, Medina et al. synthesized the COFs using a novel building block called thiophene-extended BTT. Crystalline and porous imine-linked COFs named BTT TTA and BTT TTTBA were obtained by combining BTT with triazine-based amines such as (1,3,5-triazine-2,4,6-triyl)tris(([1,1′-biphenyl]-4-amine)) (TTTBA) or 1,3,5-triazine-2,4,6-triyl trianiline (TTA). These COFs showed large surface areas, ranging between 932 and 1200 m^2 g^{-1}. The results revealed the oriented thin films which were successfully generated on a variety of substrates utilizing grazing incidence diffraction (GID) analysis. Both COFs, BTT TTTBA and BTT TTA displayed electrical conductivity up to 10^{-4} S m^{-1} at room temperature. The activation energies for BTT TTA and BTT TTTBA were, respectively, 123.3 MeV and 137.5 MeV obtained by measurements of temperature-dependent electrical conductivity.

Furthermore, charge transport in both COFs was primarily controlled by trap-assisted hopping mechanisms [195]. Moreover, Fang et al. presented the first instance of 3D tetrathiafulvalene based on COFs (3D-TTF-COFs) in a different study. These COFs showed tunable electrochemical activity with non- or 2-fold interpenetrated points structure. The synthetic COFs illustrated a large specific surface area of up to 3000 $m^2\ g^{-1}$, excellent crystallinity, and persistent porosity. These TTF-based COFs also exhibited redox activity, allowing for the iodine doping of organic salts with regulated electrical conductivity (1.4×10^{-2} S cm^{-1} at 120 °C). These results showed new design paths for 3D electroactive COF materials, highlighting the potential for these materials used in molecular electronics and energy storage [182]. Additionally, Mirica et al. described the aromatic annulation of 2,3,9,10,16,17,23,24-octa-aminophthalocyanine nickel(II) and Py-4,5,9,10-tetraone to produce a novel intrinsically conductive 2D-COF. By doping I2, the COF material (COF-DC-8) demonstrated an inherent bulk conductivity of 2.51×10^{-3} S m^{-1} that was increased by three orders of magnitude. Electronic simulations revealed an anisotropic band structure, with out-of-plane charge transport possibly making a sizable contribution to the intrinsic bulk conductivity. This conductive COF demonstrated extraordinary sensitivity to several reducing and oxidizing gases, including H_2S, NH_3, NO_2, and NO, when integrated into chemiresistive devices. Particularly, it only required a 1.5-minute exposure to obtain parts per billion (ppb) limits of detection (H_2S = 204 ppb, NH_3 = 70 ppb, NO_2 = 16 ppb, and NO = 5 ppb). Studies of spectroscopic characterization revealed that the chemiresistive response of COF-DC-8 was a result of CT interactions between the nickel phthalocyanine component of the framework and the analyte [196]. Hupp et al. also investigated three methods for improving the electrical conductivity of Zr-based MOFs. The results showed that charge was transmitted through length ranges from nanometers to micrometers. The first method essentially involved redox-hopping between sites that were chemically similar and regularly spaced apart while engaging in recurrent electron (or hole) self-exchange. Zirconium nodes were employed as grafting sites for either (a) molecular redox couples or (b) redox-active inorganic clusters, despite the fact that they were electrically insulating on their own. This alternative technique utilized charge hopping by using the redox characteristics of linkers. Marcus' CT theory has also proved effective in explaining and predicting redox hopping based on conductivity trends, particularly when it originated to account for shifts as large as 3000 folds in structurally anisotropic MOFs. Moving compensatory ions are well known as a crucial component of electronic charge propagation through redox hopping in MOF environments. As a result, the kind and quantity of permeating ions in the MOF showed effects on the rates of redox hopping. The ability to transmit electronic charge quickly enough to equal or exceed the intrinsic activity of MOF-based or MOF-immobilized catalysts is essential for electrocatalysis. Another technique for increasing electronic conductivity involves the introduction of molecular guests that can form D-A CT complexes with the host framework. This method also allows for the adjustment of conductivity using theoretical hypotheses. Adding bridges in the form of inorganic clusters or molecular oligomers that can extend the length of an MOF crystallite is a third technique that also focuses on electronic conductivity. All approaches emphasize molecular-scale porosity preservation because it is essential for numerous applications [197]. Several researchers have been motivated to improve the capabilities of pure MOFs by developing composites with metal oxides such as ZnO@ZIF-8, Fe_3O_4@MIL-100(Fe), UiO-66@WO_3, and Bi_2O_3@ HKUST-1 [176]. Bein et al. reported employing five distinct divalent metal ions, namely Mg^{2+}, Zn^{2+}, Ni^{2+}, Mn^{2+}, and Co^{2+} to introduce anthracene-based building blocks into the MOF-74 framework to investigate the electrical conductivity of MOFs. As a result, a group of extremely crystalline MOFs named ANMOF-74 (M), were created. These MOFs showed distinguished photoluminescence capabilities, with ANMOF-74(Zn) producing light throughout the whole visible spectrum. These materials illustrated enormous surface areas (1757 $m^2\ g^{-1}$) together with electrical conductivity (10^{-7} to 10^{-9} S cm^{-1}) and photoluminescence properties. In addition, a considerable increase in conductivity, up to six orders of magnitude, was observed in comparison to the original MOF-74 materials produced using 2,5-dihydroxy terephthalic acid and the same combination of metal ions (Zn^{2+}, Mg^{2+}, Ni^{2+}, Co^{2+}, and Mn^{2+}) [199]. Furthermore, Alkordi et al. presented a novel method for fabricating the composite of graphene and the HKUST-1 (MOF) used for the production of solid substrate coatings. They used commercially available, unmodified graphene to protect its electrical characteristics and prevent chemical functionalization-induced degradation (such as oxidation leading to graphene oxide, or GO). The one-pot synthesis method led to precise control over the loading of graphene and porosity tuning, offering potential application in other MOF systems. The monolithic coatings produced using this method presented high surface areas between 1156 and 1078 m^2 g^{-1}. These coatings showed high electrical conductivities that ranged from 7.6×10^{-6} S m^{-1} to 6.4×10^{-1} S m^{-1}, directly proportional to the amount of graphene present. These conductive microporous composites revealed the potential of MOF@G for a variety of applications in electrochemical catalysis, supercapacitors, and sensors [200]. Moreover, the TCNQ@Cu_3BTC_2 host-guest system (BTC = 1,3,5-benzenetricarboxylate) was produced as a remarkable example of how guest incorporation introduces semiconductivity into an originally insulating parent material. These materials showed both microporosity and semiconducting properties, presenting exciting prospects as sensor materials in the next generation. In this study, Fischer et al. developed a series of compounds with the general formula xTCNQ@Cu_3BTC_2 ($0 \leq x \leq 1.0$) by employing a solvent-free vapor phase loading technique under rigorous exclusion of moisture. They increased the specific electrical conductivity of the compound 1.0TCNQ@Cu_3BTC_2 up to 1.5×10^{-4} S cm^{-1} while keeping a high BET surface area of 573.7 m^2 g^{-1}. These characteristics made this material more desirable for uses like supercapacitors and chemoreceptors than those of other MOFs with equivalent electrical conductivity. The results provided strong evidence for the structure-property link discovered in TCNQ@ Cu_3BTC_2 and offered a strong basis for applying this design paradigm to electrically conductive MOFs [201]. Plus, Farha et al. modified the zirconium-based NU-1000 MOF by adding nickel(IV) bis(dicarbollide) to produce an electrically conductive MOF with mesoporosity. In this composite, all the nickel bis(dicarbollide) units served as guest molecules that were housed inside microporous channels of NU-1000. Other active species were introduced into the remaining mesopores of the framework thanks to this configuration. They applied manganese oxide to the electrically conductive MOF nodes. The manganese oxide contained in the conductive framework presented electrochemically accessible fractions and specific capacitance 10 times higher than that presented in the parent MOF.

NiCB@NU-100015 structure showed a surface area of 1260 $m^2\ g^{-1}$ and an electrical conductivity of 2.7 × 10^{-7} S cm^{-1} [202]. Souto et al. employed a mesoporous MOF denoted as $Fe_3O(TTFTB)_3$ (MUV-2) with a redox-active TTF-based ligand to encapsulate C_{60}, as detailed in their study. This MOF, akin to pyrene-based counterparts, exhibited a CT band in UV–vis spectroscopy, indicating D-A interactions between C_{60} and the TTF core. Despite favorable ligand-to-guest CT and a reduced bandgap predicted by DFT calculations for the C_{60}-encapsulated material, the observed increase in conductivity was modest (from 3.7 × 10^{-11} S cm^{-1} for MUV-2 to 4.7 × 10^{-9} S cm^{-1} for C_{60}@MUV-2). The authors attributed this relatively low conductivity value to the low loading of C_{60} (C_{60}@MUV-2 maintained high porosity with a BET surface area of 1040 $m^2\ g^{-1}$) and the considerable distances between the ligand TTF moieties [203]. Dincă et al. especially investigated DMF coordination with the open Fe sites, leading to a three-orders-of-magnitude improvement in electrical conductivity and a surface area of 83 $m^2\ g^{-1}$ at room temperature. The open Fe centers in Fe_2(DSBDC) (where DSBDC4- = 2,5-disulfidobenzene-1,4-dicarboxylate) experienced inner-sphere modifications, leading to reversible changes in structure and electrical conductivity. Additional DFT calculations were performed, and the results showed that the increase in electrical conductivity (3.9 × 10^{-6} S cm^{-1}) resulted from a partial electron transfer from Fe to DMF. Plus, Fe_2(DSBDC) showed a greater charge carrier density because of the formation of hole carriers [204]. Pure CMP electrodes often have weak electrical conductivity, which limits the use of the material in electrochemical applications. The electrical conductivity of CMPs is determined by several factors, including the type and length of the conjugated backbone, the degree of doping, and the presence of defects in the polymer structure. Generally, CMPs exhibited low electrical conductivity in their pristine state due to their intrinsic insulating nature. However, doping CMPs with electron-rich or electron-deficient dopants can enhance their electrical conductivity [205]. Qiao et al. synthesized the 2D laminar covalent quinazoline network (TQ-CQN) and 3D amorphous CMP (TQ-CMP) by tuning the geometry of the N-rich skeleton tricycloquinazoline (TQ) unit. In terms of electrochemical catalysis, TQ-CQN exhibited a lower CT resistance than TQ-CMP [206]. Such porous materials offer distinct advantages, including immense structural and functional versatility. They combine porosity with diverse electronic properties and can have their physical attributes fine-tuned through post-synthetic modifications [193].

Generally, design principles of CT based on CMPs and COFs include incorporation of π-conjugated systems to enhance intrinsic electrical conductivity, engineering D-A motifs within COF structures to facilitate charge separation and transfer, and designing redox-active sites for efficient charge storage and transport. In MOFs, it is based on introducing metal centers with variable oxidation states to enable redox chemistry for CT processes, utilizing ligands with suitable redox potentials to facilitate electron transfer reactions, and designing MOF structures with open metal sites to enhance charge mobility and facilitate CT. In this case, there are some main challenges such as synthetic control, stability, scalability, and understanding CT mechanisms. Achieving precise control over the synthesis of COFs and MOFs, especially concerning their crystallinity and defect formation, remains a significant challenge. Ensuring stability under various environmental conditions, especially for MOFs, is critical for practical applications. Strategies to enhance stability without compromising their intrinsic properties are actively being researched. Developing scalable and cost-effective synthesis methods for both COFs and MOFs to facilitate their large-scale production while maintaining the desired properties. Further understanding of the CT mechanisms within these materials is necessary to optimize their design for specific applications, such as energy storage, catalysis, and electronic devices. Notably, a challenge within this area is the engineering of framework materials with desired redox properties targeted at a specific function [35c].

7.8 Prospect for CT Complex

Here is an exhaustive discussion of the advantages and disadvantages of CT in the three different types of materials (CMPs, COFs, and MOFs). CMPs offer intriguing opportunities for CT complexes due to their complex structures and diverse chemical functionalities. The presence of multiple components in CMPs allows for the incorporation of different electron-D and A moieties, facilitating CT interactions. This feature can be exploited to design CMPs with tailored electronic properties and to promote CT processes within the material. It is possible to create CMPs that improve CT efficiency by carefully selecting the building blocks, which can be useful for applications such as energy storage, electrochromic devices, and sensors. Furthermore, COFs, being organic-based porous materials, also have the potential for the formation of CT complexes. Their crystalline nature and the ability to incorporate different organic linkers enable the accurate positioning of electron-D and A groups, facilitating CT interactions. The design of COFs with specific functionalities can promote efficient CT and facilitate applications such as optoelectronic devices, photovoltaics, and sensing platforms. However, it is worth noting that the synthesis of COFs with precise control over CT properties can be challenging, and further research is needed to optimize their design and performance. In addition, MOFs are known for their high tunability and various metal and organic linker combinations, making them attractive for CT complex applications. The presence of metal centers in MOFs can facilitate CT processes by serving as electron-D and A. Moreover, the presence of well-defined pore structures in MOFs can enhance CT efficiency by promoting the diffusion and transport of charge carriers. These properties make MOFs potential candidates for applications such as photocatalysis, energy conversion and storage, and electrochemical sensing. However, the selection of appropriate metal ions and organic linkers is crucial to optimize the CT behavior of MOFs. On the other hand, in COFs and MOFs, the crystalline order refers to the well-defined, periodic arrangement of covalently bonded organic units. This ordered structure creates continuous pathways for charge carriers, allowing for efficient CT. Crystallinity in COFs and MOFs is crucial for achieving high electrical conductivity because it minimizes defects and disruptions in the charge transport pathways. Highly crystalline exhibits enhanced electrical conductivity and improved CT kinetics, making them promising materials for various electronic and optoelectronic applications. CMPs, on the other hand, often possess a more amorphous or disordered structure compared to COFs. The lack of long-range order in CMPs can result in limited and discontinuous charge transport pathways. While some CMPs can exhibit semi-crystalline regions, the overall impact of crystallinity on electrical conductivity is less pronounced than in highly crystalline COFs. General principles for the design of CT in each of these materials could also be highlighted as well as the main challenges.

In terms of prediction, it is expected that further advancements will be made in designing and tailoring the porous frameworks for CT complex applications. Enhancing CT performance and creating new possibilities for diverse applications will depend heavily on the capacity to accurately manipulate the arrangement of electron-D and A groups within these materials. On the contrary, a serious problem was created in the form of electron-hole recombination when the stacking of D and A units in COFs was disordered. This recombination process was not desirable for specific applications such as solar cells and field-effect transistors. Consequently, the lack of CT pathways or effective separation of charge carriers has been recognized as a challenge in COFs. However, the distinctive structural features of D-A COFs offer potential solutions by providing pathways and channels for the efficient transport of charge carriers [151k]. However, it is important to note that challenges related to material synthesis, stability, and scalability need to be addressed for the widespread implementation of CT complexes in CMPs, COFs, and MOFs.

7.9 Conclusion

In conclusion, this review article provides a broad overview of CT phenomena in CMP, COF, and MOF structures. Throughout the article, we have discussed various D-A structures in CMP, COF, and MOF for semiconductor applications. The study of CT processes in these materials is of great importance due to their potential applications in various fields such as energy storage, catalysis, sensing, and electronics. In this paper, various strategies were studied to regulate the CT of these materials, including the introduction of electron-donating and electron-accepting groups, the incorporation of redox-active sites, and the use of guest molecules. These approaches have enabled researchers to tune the electronic properties of CMP, COF, and MOF to achieve improved conductivity, enhanced catalytic activity, and efficient charge storage functions. In addition, the review introduces a variety of experimental and theoretical studies employed to investigate CT of these porous frameworks. Experimental methods such as spectroscopy, electrochemistry, and conductivity measurements have provided valuable insights into the mechanism and kinetics of CT. On the other hand, theoretical calculations such as DFT calculations have provided a deeper understanding of the electronic structure and charge-transport properties of these materials. The field and semiconductor applications have come a long way, but challenges remain. More research is needed to understand the true mechanisms governing CT of these materials and to develop more accurate theoretical models to guide the prediction and design of new functional materials. In addition, the scalability and stability of CMP, COF, and MOF will also need to be addressed to ensure real-world practicality. In conclusion, the study of CT in CMP, COF, and MOF shows surprising potential for the development of novel functional materials. By unraveling the underlying principles and exploring creative methods, researchers can use CT phenomena to create advanced materials with quality tailored to various applications. This will advance technology in a variety of industries.

References

1. (a) K. Geng, T. He, R. Liu, S. Dalapati, K. T. Tan, Z. Li, S. Tao, Y. Gong, Q. Jiang and D. Jiang, *Chemical Reviews* 2020, 120, 8814–8933.

   (b) R.-B. Lin, Y. He, P. Li, H. Wang, W. Zhou and B. Chen, *Chemical Society Reviews* 2019, 48, 1362–1389.

   (c) R.-R. Liang, S.-Y. Jiang, A. Ru-Han and X. Zhao, *Chemical Society Reviews* 2020, 49, 3920–3951.

   (d) D. Wang, G. Guo, R. Chen, Y. Gong, L. Sun and Y. Zhao, *Science China Chemistry* 2021, 64, 1510–1514.

2. H. Veldhuizen, L. van Den Elzen, T. Mahon, R. Abellon and A. Nagai, *Macromolecular Chemistry and Physics* 2020, 221, 1900415.

3. (a) G. Lin, H. Ding, R. Chen, Z. Peng, B. Wang and C. Wang, *Journal of the American Chemical Society* 2017, 139, 8705–8709.

   (b) X. Huang, C. Sun and X. Feng, *Science China Chemistry* 2020, 63, 1367–1390.

4. (a) H. Furukawa, K. E. Cordova, M. O'Keeffe and O. M. Yaghi, *Science* 2013, 341, 1230444.

   (b) T. D. Duong, S. A. Sapchenko, I. da Silva, H. G. Godfrey, Y. Cheng, L. L. Daemen, P. Manuel, M. D. Frogley, G. Cinque and A. J. Ramirez-Cuesta, *Chemical Science* 2020, 11, 5339–5346.

5. (a) Y. Xu, S. Jin, H. Xu, A. Nagai and D. Jiang, *Chemical Society Reviews* 2013, 42, 8012–8031.

(b) S. Luo, Z. Zeng, H. Wang, W. Xiong, B. Song, C. Zhou, A. Duan, X. Tan, Q. He and G. Zeng, *Progress in Polymer Science* 2021, 115, 101374.

6. A. Nagai, *Covalent Organic Frameworks*, CRC Press, 2019.

7. M. Shahmirzaee, J. Abdi, A. Hemmati-Sarapardeh, M. Schaffie, M. Ranjbar and A. Khataee, *Journal of Molecular Liquids* 2022, 363, 119900.

8. M. Dogru, M. Handloser, F. Auras, T. Kunz, D. Medina, A. Hartschuh, P. Knochel and T. Bein, *Angewandte Chemie International Edition* 2013, 52, 2920–2924.

9. Z. J. Wang, S. Ghasimi, K. Landfester and K. A. Zhang, *Advanced Materials* 2015, 27, 6265–6270.

10. (a) J. L. Segura, M. J. Mancheño and F. Zamora, *Chemical Society Reviews* 2016, 45, 5635–5671.

(b) Y. Li, W. Chen, G. Xing, D. Jiang and L. Chen, *Chemical Society Reviews* 2020, 49, 2852–2868.

11. M. B. Solomon, T. L. Church and D. M. D'Alessandro, *CrystEngComm* 2017, 19, 4049–4065.

12. K. Jayaramulu, S. K. Reddy, A. Hazra, S. Balasubramanian and T. K. Maji, *Inorganic Chemistry* 2012, 51, 7103–7111.

13. (a) C.-W. Kung, S. Goswami, I. Hod, T. C. Wang, J. Duan, O. K. Farha and J. T. Hupp, *Accounts of Chemical Research* 2020, 53, 1187–1195.

(b) N. D. Shepherd and D. M. D'Alessandro, *Chemical Physics Reviews* 2021, 2.

14. S. Li, Z.-D. Yang, G. Zhang and X. C. Zeng, *Journal of Materials Chemistry C* 2015, 3, 9637–9649.

15. (a) S. Iwai, S. Murata and M. Tachiya, *The Journal of Chemical Physics* 1998, 109, 5963–5970.

(b) J. R. Winkler, T. L. Netzel, C. Creutz and N. Sutin, *Journal of the American Chemical Society* 1987, 109, 2381–2392.

16. A. Kuznetsova, V. Matveevskaya, D. Pavlov, A. Yakunenkov and A. Potapov, *Materials* 2020, 13, 2699.

17. V. Martínez-Martínez, S. Furukawa, Y. Takashima, I. Lopez Arbeloa and S. Kitagawa, *The Journal of Physical Chemistry C* 2012, 116, 26084–26090.

18. R. Haldar, R. Matsuda, S. Kitagawa, S. J. George and T. K. Maji, *Angewandte Chemie International Edition* 2014, 53, 11772–11777.

19. L. Ascherl, T. Sick, J. T. Margraf, S. H. Lapidus, M. Calik, C. Hettstedt, K. Karaghiosoff, M. Döblinger, T. Clark, K. W. Chapman, F. Auras and T. Bein, *Nat. Chem.* 2016, 8, 310.
20. N. Sikdar, A. Hazra, D. Samanta, R. Haldar and T. K. Maji, *Angewandte Chemie* 2020, 132, 18637–18642.
21. M. Li, C. An, W. Pisula and K. Müllen, *Accounts of Chemical Research* 2018, 51, 1196–1205.
22. S. Gunes, H. S. Neugebauer and N. S. Sariciftci, *Chem. Rev.* 2007, 107, 1324.
23. R. C. Coffin, J. Peet, J. Rogers and G. C. Bazan, *Nat. Chem.* 2009, 1, 657.
24. Y. Zhang, Y. Xu, M. J. Ford, F. Li, J. Sun, X. Ling, Y. Wang, J. Gu, J. Yuan and W. Ma, *Advanced Energy Materials* 2018, 8, 1800029.
25. N. Van den Brande, N. Patil, M. Guizar-Sicairos, R. Claessens, G. Van Assche, D. W. Breiby and B. Van Mele, *Organic Electronics* 2017, 41, 319–326.
26. H.-Y. Chen, J. Hou, S. Zhang, Y. Liang, G. Yang, Y. Yang, L. Yu, Y. Wu and G. Li, *Nature Photonics* 2009, 3, 649–653.
27. L. Yao, Y. Zhang, H.-X. Wang, Y. Guo, Z.-M. Zhuang, W. Wen, X. Zhang and S. Wang, *Journal of Materials Chemistry A* 2020, 8, 8518–8526.
28. Y. Yang, Z. Liu, G. Zhang, X. Zhang and D. Zhang, *Advanced Materials* 2019, 31, 1903104.
29. F. Wudl, D. Wobschall and E. J. Hufnagel, *Journal of the American Chemical Society* 1972, 94, 670–672.
30. (a) J. Ferraris, D. Cowan, V. T. Walatka and J. Perlstein, *Journal of the American Chemical Society* 1973, 95, 948–949.

    (b) M. Bendikov, F. Wudl and D. F. Perepichka, *Chemical Reviews* 2004, 104, 4891–4946.
31. M.-K. Ng, D.-C. Lee and L. Yu, *Journal of the American Chemical Society* 2002, 124, 11862–11863.
32. R. M. Metzger, *Chemical Reviews* 2003, 103, 3803–3834.
33. A. A. Kocherzhenko, S. Patwardhan, F. C. Grozema, H. L. Anderson and L. D. Siebbeles, *Journal of the American Chemical Society* 2009, 131, 5522–5529.
34. (a) S. S. Park, E. R. Hontz, L. Sun, C. H. Hendon, A. Walsh, T. Van Voorhis and M. Dincă, *Journal of the American Chemical Society* 2015, 137, 1774–1777.

    (b) S. R. Ahrenholtz, C. C. Epley and A. J. Morris, *Journal of the American Chemical Society* 2014, 136, 2464–2472.

(c) S. Patwardhan and G. C. Schatz, *The Journal of Physical Chemistry C* 2015, 119, 24238–24247.

35. (a) M. Cai, Q. Loague and A. J. Morris, *The Journal of Physical Chemistry Letters* 2020, 11, 702–709.

    (b) A. Dey, F. A. Rahimi, P. Verma, S. Suresh and T. K. Maji, *ACS Applied Energy Materials* 2023, 6, 9179–9187.

    (c) B. Ding, M. B. Solomon, C. F. Leong and D. M. D'Alessandro, *Coordination Chemistry Reviews* 2021, 439, 213891.

36. C. Hua, P. W. Doheny, B. Ding, B. Chan, M. Yu, C. J. Kepert and D. M. D'Alessandro, *Journal of the American Chemical Society* 2018, 140, 6622–6630.

37. K. P. Goetz, D. Vermeulen, M. E. Payne, C. Kloc, L. E. McNeil and O. D. Jurchescu, *Journal of Materials Chemistry C* 2014, 2, 3065–3076.

38. H. Miyasaka, *Accounts of Chemical Research* 2013, 46, 248–257.

39. N. Chaoui, M. Trunk, R. Dawson, J. Schmidt and A. Thomas, *Chemical Society Reviews* 2017, 46, 3302–3321.

40. Y. Liao, H. Wang, M. Zhu and A. Thomas, *Advanced Materials* 2018, 30, 1705710.

41. X. Zhu, C. Tian, G. M. Veith, C. W. Abney, J. Dehaudt and S. Dai, *Journal of the American Chemical Society* 2016, 138, 11497–11500.

42. T. Geng, Z. Zhu, W. Zhang and Y. Wang, *Journal of Materials Chemistry A* 2017, 5, 7612–7617.

43. G. Dong, L. Yang, F. Wang, L. Zang and C. Wang, *ACS Catalysis* 2016, 6, 6511–6519.

44. (a) M. Thommes, K. Kaneko, A. V. Neimark, J. P. Olivier, F. Rodriguez-Reinoso, J. Rouquerol and K. S. W. Sing, *Pure and Applied Chemistry* 2015, 87, 1051–1069.

    (b) H. Ma, Y. Chen, X. Li and B. Li, *Advanced Functional Materials* 2021, 31, 2101861.

45. J.-S. M. Lee and A. I. Cooper, *Chemical Reviews* 2020, 120, 2171–2214.

46. S. Yang, C. Yang, X. Zhang, Z. Zheng, S. Bi, Y. Zhang and H. Zhou, *Journal of Materials Chemistry C* 2018, 6, 9044–9048.

47. (a) S. Ren, M. J. Bojdys, R. Dawson, A. Laybourn, Y. Z. Khimyak, D. J. Adams and A. I. Cooper, *Advanced Materials* 2012, 24, 2357–2361.

    (b) P. Katekomol, J. Roeser, M. Bojdys, J. Weber and A. Thomas, *Chem. Mater.* 2013, 25, 1542.

48. P. Kuhn, M. Antonietti and A. Thomas, *Angewandte Chemie International Edition* 2008, 47, 3450–3453.

49. J. Chen, W. Yan, E. J. Townsend, J. Feng, L. Pan, V. Del Angel Hernandez and C. F. Faul, *Angewandte Chemie International Edition* 2019, 58, 11715–11719.

49. Y. Xu, L. Chen, Z. Guo, A. Nagai and D. Jiang, *J. Am. Chem. Soc.* 2011, 133, 17622.

50. Y. Xu, A. Nagai and D. Jiang, *Chemical Communications* 2013, 49, 1591–1593.

51. S. N. Talapaneni, D. Kim, G. Barin, O. Buyukcakir, S. H. Je and A. Coskun, *Chem. Mater.* 2016, 28, 4460.

52. Y. Kou, Y. Xu, Z. Guo and D. Jiang, *Angewandte Chemie* 2011, 123, 8912–8916.

53. A. F. El-Mahdy, C.-H. Kuo, A. Alshehri, C. Young, Y. Yamauchi, J. Kim and S.-W. Kuo, *Journal of Materials Chemistry A* 2018, 6, 19532–19541.

55. A. F. M. El-Mahdy, C. Young, J. Kim, J. You, Y. Yamauchi and S.-W. Kuo, *ACS Applied Materials & Interfaces* 2019, 11, 9343–9354.

56. (a) Y. Miseki and K. Sayama, *Advanced Energy Materials* 2019, 9, 1801294.

(b) M.-Q. Yang, M. Gao, M. Hong and G. W. Ho, *Advanced Materials* 2018, 30, 1802894.

(c) J. Fu, Q. Xu, J. Low, C. Jiang and J. Yu, *Applied Catalysis B: Environmental* 2019, 243, 556–565.

57. F. J. Rizzuto, T. B. Faust, B. Chan, C. Hua, D. M. D'Alessandro and C. J. Kepert, *Chemistry – A European Journal* 2014, 20, 17597–17605.

58. A. F. Saber, A. M. Elewa, H.-H. Chou and A. F. EL-Mahdy, *Applied Catalysis B: Environmental* 2022, 316, 121624.

59. Z. Li, H. Li, H. Xia, X. Ding, X. Luo, X. Liu and Y. Mu, *Chemistry–A European Journal* 2015, 21, 17355–17362.

60. L. Li, Z. Cai, Q. Wu, W.-Y. Lo, N. Zhang, L. X. Chen and L. Yu, *Journal of the American Chemical Society* 2016, 138, 7681–7686.

61. L. Li, W.-y. Lo, Z. Cai, N. Zhang and L. Yu, *Macromolecules* 2016, 49, 6903–6909.

62. Y. Z. Peng, G. C. Guo, S. Guo, L. H. Kong, T. B. Lu and Z. M. Zhang, *Angewandte Chemie* 2021, 133, 22233–22240.

63. A. Nagai and Y. Okabe, *Chemical Communications* 2014, 50, 10052–10054.

64. Y. Xu, N. Mao, C. Zhang, X. Wang, J. Zeng, Y. Chen, F. Wang and J.-X. Jiang, *Applied Catalysis B: Environmental* 2018, 228, 1–9.

65. Y. S. Kochergin, D. Schwarz, A. Acharjya, A. Ichangi, R. Kulkarni, P. Eliášová, J. Vacek, J. Schmidt, A. Thomas and M. J. Bojdys, *Angewandte Chemie International Edition* 2018, 57, 14188–14192.

66. V. S. Mothika, P. Sutar, P. Verma, S. Das, S. K. Pati and T. K. Maji, *Chemistry–A European Journal* 2019, 25, 3867–3874.

67. S. Roy, A. Bandyopadhyay, M. Das, P. P. Ray, S. K. Pati and T. K. Maji, *Journal of Materials Chemistry A* 2018, 6, 5587–5591.

68. V. S. Mothika, P. Sutar, P. Verma, S. Das, S. K. Pati and T. K. Maji, *Chem. - Eur. J.* 2019, 25, 3867.

69. H. Bildirir, I. Osken, T. Ozturk and A. Thomas, *Chemistry–A European Journal* 2015, 21, 9306–9311.

70. C. Gu, N. Huang, Y. Chen, L. Qin, H. Xu, S. Zhang, F. Li, Y. Ma and D. Jiang, *Angew. Chem., Int. Ed.* 2015, 54, 13594.

71. C. Gu, N. Huang, Y. Chen, L. Qin, H. Xu, S. Zhang, F. Li, Y. Ma and D. Jiang, *Angew. Chem.* 2015, 127, 13798.

72. C. Gu, N. Huang, Y. Chen, L. Qin, H. Xu, S. Zhang, F. Li, Y. Ma and D. Jiang, *Angewandte Chemie* 2015, 127, 13798–13802.

73. X. Feng, L. Liu, Y. Honsho, A. Saeki, S. Seki, S. Irle, Y. Dong, A. Nagai and D. Jiang, *Angewandte Chemie* 2012, 124, 2672–2676.

74. X. Guan, Y. Ma, H. Li, Y. Yusran, M. Xue, Q. Fang, Y. Yan, V. Valtchev and S. Qiu, *Journal of the American Chemical Society* 2018, 140, 4494–4498.

75. T. Li, X. Yan, W.-D. Zhang, W.-K. Han, Y. Liu, Y. Li, H. Zhu, Z. Li and Z.-G. Gu, *Chemical Communications* 2020, 56, 14187–14190.

76. M. Dogru, M. Handloser, F. Auras, T. Kunz, D. Medina, A. Hartschuh, P. Knochel and T. Bein, *Angew. Chem., Int. Ed.* 2013, 52, 2920.

77. Z. Li, Z. Zhang, R. Nie, C. Li, Q. Sun, W. Shi, W. Chu, Y. Long, H. Li and X. Liu, *Advanced Functional Materials* 2022, 32, 2112553.

78. X. Feng, L. Chen, Y. Honsho, O. Saengsawang, L. Liu, L. Wang, A. Saeki, S. Irle, S. Seki and Y. Dong, *Advanced Materials* 2012, 24, 3026–3031.

79. S. Jin, K. Furukawa, M. Addicoat, L. Chen, S. Takahashi, S. Irle, T. Nakamura and D. Jiang, *Chemical Science* 2013, 4, 4505–4511.

80. J. Guo, Y. Xu, S. Jin, L. Chen, T. Kaji, Y. Honsho, M. A. Addicoat, J. Kim, A. Saeki and H. Ihee, *Nat. Commun.* 2013, 4, 2736.

81. A. M. Rice, E. A. Dolgopolova, B. J. Yarbrough, G. A. Leith, C. R. Martin, K. S. Stephenson, R. A. Heugh, A. J. Brandt, D. A. Chen and S. G. Karakalos, *Angewandte Chemie International Edition* 2018, 57, 11310–11315.

82. G. A. Leith, A. M. Rice, B. J. Yarbrough, A. A. Berseneva, R. T. Ly, C. N. Buck III, D. Chusov, A. J. Brandt, D. A. Chen and B. W. Lamm, *Angewandte Chemie* 2020, 132, 6056–6062.

83. S. Jin, M. Supur, M. Addicoat, K. Furukawa, L. Chen, T. Nakamura, S. Fukuzumi, S. Irle and D. Jiang, *J. Am. Chem. Soc.* 2015, 137, 7817.

84. T. W. Kim, S. Jun, Y. Ha, R. K. Yadav, A. Kumar, C.-Y. Yoo, I. Oh, H.-K. Lim, J. W. Shin and R. Ryoo, *Nature Communications* 2019, 10, 1873.

85. L. Chen, K. Furukawa, J. Gao, A. Nagai, T. Nakamura, Y. Dong and D. Jiang, *Journal of the American Chemical Society* 2014, 136, 9806–9809.

86. S. Jin, X. Ding, X. Feng, M. Supur, K. Furukawa, S. Takahashi, M. Addicoat, M. E. El-Khouly, T. Nakamura and S. Irle, *Angewandte Chemie International Edition* 2013, 7, 2017–2021.

87. X. Chen, M. Addicoat, S. Irle, A. Nagai and D. Jiang, *J. Am. Chem. Soc.* 2012, 135, 546.

88. F. Auras, L. Ascherl, A. H. Hakimioun, J. T. Margraf, F. C. Hanusch, S. Reuter, D. Bessinger, M. Döblinger, C. Hettstedt and K. Karaghiosoff, *Journal of the American Chemical Society* 2016, 138, 16703–16710.

89. H. Ding, Y. Li, H. Hu, Y. Sun, J. Wang, C. Wang, C. Wang, G. Zhang, B. Wang and W. Xu, *Chemistry–A European Journal* 2014, 20, 14614–14618.

90. C. Krishnaraj, H. Sekhar Jena, L. Bourda, A. Laemont, P. Pachfule, J. r. m. Roeser, C. V. Chandran, S. Borgmans, S. M. Rogge and K. Leus, *Journal of the American Chemical Society* 2020, 142, 20107–20116.

91. R. Luo, H. Lv, Q. Liao, N. Wang, J. Yang, Y. Li, K. Xi, X. Wu, H. Ju and J. Lei, *Nature Communications* 2021, 12, 6808.

92. D. Wang, H. Zeng, X. Xiong, M.-F. Wu, M. Xia, M. Xie, J.-P. Zou and S.-L. Luo, *Science Bulletin* 2020, 65, 113–122.

93. T. Joshi, C. Chen, H. Li, C. S. Diercks, G. Wang, P. J. Waller, H. Li, J. L. Bredas, O. M. Yaghi and M. F. Crommie, *Advanced Materials* 2019, 31, 1805941.

94. Y. Zhong, B. Cheng, C. Park, A. Ray, S. Brown, F. Mujid, J.-U. Lee, H. Zhou, J. Suh and K.-H. Lee, *Science* 2019, 366, 1379–1384.

95. S. T. Meek, J. A. Greathouse and M. D. Allendorf, *Advanced Materials* 2011, 23, 249–267.

96. (a) H. Meng, Y. Han, C. Zhou, Q. Jiang, X. Shi, C. Zhan and R. Zhang, *Small Methods* 2020, 4, 2000396.

(b) C. G. Silva, A. Corma and H. García, *Journal of Materials Chemistry* 2010, 20, 3141–3156.

97. (a) I. Hod, O. K. Farha and J. T. Hupp, *Chem. Commun.* 2016, 52, 1705.

(b) C.-W. Kung, T.-H. Chang, L.-Y. Chou, J. T. Hupp, O. K. Farha and K.-C. Ho, *Electrochemistry Communications* 2015, 58, 51–56.

98. S. K. Park, S. Varghese, J. H. Kim, S.-J. Yoon, O. K. Kwon, B.-K. An, J. Gierschner and S. Y. Park, *Journal of the American Chemical Society* 2013, 135, 4757–4764.

99. N. Hush, *Electrochim. Acta* 1968, 13, 1005.

100. X. Zhang, I. Da Silva, R. Fazzi, A. M. Sheveleva, X. Han, B. F. Spencer, S. A. Sapchenko, F. Tuna, E. J. McInnes and M. Li, *Inorganic Chemistry* 2019, 58, 14145–14150.

101. J. Yang, L. Zhou, J. Cheng, Z. Hu, C. Kuo, C.-W. Pao, L. Jang, J.-F. Lee, J. Dai and S. Zhang, *Inorganic Chemistry* 2015, 54, 6433–6438.

102. T. Tachikawa, J. R. Choi, M. Fujitsuka and T. Majima, *The Journal of Physical Chemistry C* 2008, 112, 14090–14101.

103. P. Majee, D. K. Singha, S. K. Mondal and P. Mahata, *Photochemical & Photobiological Sciences* 2019, 18, 1110–1121.

104. X.-Y. Xu and B. Yan, *Journal of Materials Chemistry A* 2017, 5, 2215–2223.

105. X. Li, J. Yu, D. J. Gosztola, H. C. Fry and P. Deria, *Journal of the American Chemical Society* 2019, 141, 16849–16857.

106. C. Xu, Y. Pan, G. Wan, H. Liu, L. Wang, H. Zhou, S.-H. Yu and H.-L. Jiang, *Journal of the American Chemical Society* 2019, 141, 19110–19117.

107. Q. Mu, Y. Su, Z. Wei, H. Sun, Y. Lian, Y. Dong, P. Qi, Z. Deng and Y. Peng, *Journal of Catalysis* 2021, 397, 128–136.

108. Z. Qi and Y. Chen, *Biosensors and Bioelectronics* 2017, 87, 236–241.

109. Y. Li, Y. Xia, K. Liu, K. Ye, Q. Wang, S. Zhang, Y. Huang and H. Liu, ACS *Applied Materials & Interfaces* 2020, 12, 25494–25502.

110. Y. Liu, J. Lu, Q. Zhang, Y. Bai, X. Pang, S. Wang, H. Bai and W. Fan, *Chemical Communications* 2021, 57, 8031–8034.

111. (a) D.-C. Liu, H.-J. Wang, J.-W. Wang, D.-C. Zhong, L. Jiang and T.-B. Lu, *Chemical Communications* 2018, 54, 11308–11311.

(b) C. Gao, J. Wang, H. Xu and Y. Xiong, *Chemical Society Reviews* 2017, 46, 2799–2823.

112. A. Crake, K. C. Christoforidis, A. Gregg, B. Moss, A. Kafizas and C. Petit, *Small* 2019, 15, 1805473.

113. Y. Zhao, J. Wang, W. Zhu, L. Liu and R. Pei, *Nanoscale* 2021, 13, 4505–4511.

114. Q. Chen, M. Luo, P. Hammershøj, D. Zhou, Y. Han, B. W. Laursen, C.-G. Yan and B.-H. Han, *Journal of the American Chemical Society* 2012, 134, 6084–6087.

115. R. Bhosale, J. Míšek, N. Sakai and S. Matile, *Chemical Society Reviews* 2010, 39, 138–149.

116. H. Y. Au-Yeung, G. D. Pantoş and J. K. Sanders, *Proceedings of the National Academy of Sciences* 2009, 106, 10466–10470.

117. P. Talukdar, G. Bollot, J. Mareda, N. Sakai and S. Matile, *Journal of the American Chemical Society* 2005, 127, 6528–6529.

118. W. Pisula, M. Kastler, D. Wasserfallen, J. W. Robertson, F. Nolde, C. Kohl and K. Müllen, *Angewandte Chemie International Edition* 2006, 45, 819–823.

119. (a) V. J. Bradford and B. L. Iverson, *Journal of the American Chemical Society* 2008, 130, 1517–1524.

(b) S. De and S. Ramakrishnan, *Macromolecules* 2009, 42, 8599–8603.

(c) J. Li, L. Guo, L. Zhang, C. Yu, L. Yu, P. Jiang, C. Wei, F. Qin and J. Shi, *Dalton Transactions* 2009, 823–831.

(d) H. Shao, T. Nguyen, N. C. Romano, D. A. Modarelli and J. R. Parquette, *Journal of the American Chemical Society* 2009, 131, 16374–16376.

(e) K. V. Rao, K. Jayaramulu, T. K. Maji and S. J. George, *Angewandte Chemie* 2010, 122, 4314–4318.

120. (a) X. Liu, Y. Xu and D. Jiang, *J. Am. Chem. Soc.* 2012, 134, 8738.

(b) W. Zhao, X. Zhuang, D. Wu, F. Zhang, D. Gehrig, F. Laquai and X. Feng, *Journal of Materials Chemistry A* 2013, 1, 13878–13884.

(c) L. Chen, Y. Honsho, S. Seki and D. Jiang, *J. Am. Chem. Soc.* 2010, 132, 6742.

121. (a) Q. Chen, D.-P. Liu, J.-H. Zhu and B.-H. Han, *Macromolecules* 2014, 47, 5926–5931.

(b) C. Zhang, Y. Liu, B. Li, B. Tan, C.-F. Chen, H.-B. Xu and X.-L. Yang, *ACS Macro Letters* 2012, 1, 190–193.

(c) P. Zhang, K. Wu, J. Guo and C. Wang, *ACS Macro Lett.* 2014, 3, 1139.

(d) V. M. Suresh, S. Bonakala, S. Roy, S. Balasubramanian and T. K. Maji, *The Journal of Physical Chemistry C* 2014, 118, 24369–24376.

122. (a) Y.-C. Zhao, T. Wang, L.-M. Zhang, Y. Cui and B.-H. Han, *Polymer Chemistry* 2015, 6, 748–753.

(b) C. Zhang, L.-H. Peng, B. Li, Y. Liu, P.-C. Zhu, Z. Wang, D.-H. Zhan, B. Tan, X.-L. Yang and H.-B. Xu, *Polymer Chemistry* 2013, 4, 3663–3666.

123. N. Huang, Y. Xu and D. Jiang, *Scientific Reports* 2014, 4, 1–8.

124. C. Xu and N. Hedin, *J. Mater. Chem. A* 2013, 1, 3406.

125. S. Xu, K. Song, T. Li and B. Tan, *Journal of Materials Chemistry A* 2015, 3, 1272–1278.

126. (a) A. I. Cooper, *Advanced Materials* 2009, 21, 1291–1295.

(b) C. Gu, Y. Chen, Z. Zhang, S. Xue, S. Sun, K. Zhang, C. Zhong, H. Zhang, Y. Pan and Y. Lv, *Advanced Materials* 2013, 25, 3443–3448.

127. J. Weber and A. Thomas, *J. Am. Chem. Soc.* 2008, 130, 6334.

128. J. Schmidt, M. Werner and A. Thomas, *Macromolecules* 2009, 42, 4426.

129. J. X. Jiang, F. Su, A. Trewin, C. D. Wood, N. L. Campbell, H. Niu, C. Dickinson, A. Y. Ganin, M. J. Rosseinsky and Y. Z. Khimyak, *Angew. Chem., Int. Ed.* 2007, 46, 8574.

130. G. Cheng, T. Hasell, A. Trewin, D. J. Adams and A. I. Cooper, *Angew. Chem., Int. Ed.* 2012, 51, 12727.

131. M. Kim, M. Byeon, J.-S. Bae, S.-Y. Moon, G. Yu, K. Shin, F. Basarir, T.-H. Yoon and J.-W. Park, *Macromolecules* 2011, 44, 7092–7095.

132. V. Senkovskyy, I. Senkovska and A. Kiriy, *ACS Macro Letters* 2012, 1, 494–498.

133. T. M. Figueira-Duarte, S. C. Simon, M. Wagner, S. I. Druzhinin, K. A. Zachariasse and K. Müllen, *Angewandte Chemie International Edition* 2008, 47, 10175–10178.

134. A. Trewin and A. I. Cooper, *Angew. Chem., Int. Ed.* 2010, 49, 1533.

135. J.-X. Jiang, A. Laybourn, R. Clowes, Y. Z. Khimyak, J. Bacsa, S. J. Higgins, D. J. Adams and A. I. Cooper, *Macromolecules* 2010, 43, 7577–7582.

136. J. Schmidt, J. Weber, J. D. Epping, M. Antonietti and A. Thomas, *Advanced Materials* 2009, 21, 702–705.

137. J. Wang, Z. Chang, B. Ding, T. Li, G. Yang, Z. Pang, T. Nakato, M. Eguchi, Y. M. Kang and J. Na, *Angewandte Chemie* 2020, 132, 19738–19743.

138. A. Rose, Z. Zhu, C. F. Madigan, T. M. Swager and V. Bulović, *Nature* 2005, 434, 876–879.

139. S. S. Han, H. Furukawa, O. M. Yaghi and W. A. Goddard Iii, *Journal of the American Chemical Society* 2008, 130, 11580–11581.

140. J. R. Hunt, C. J. Doonan, J. D. LeVangie, A. P. Côté and O. M. Yaghi, *Journal of the American Chemical Society* 2008, 130, 11872–11873.

141. M. Liu, Q. Huang, S. Wang, Z. Li, B. Li, S. Jin and B. Tan, *Angew. Chem., Int. Ed.* 2018, 57, 11968.

142. (a) F. J. Uribe-Romo, J. R. Hunt, H. Furukawa, C. Klock, M. O'Keeffe and O. M. Yaghi, *Journal of the American Chemical Society* 2009, 131, 4570–4571.

(b) T. Ma, E. A. Kapustin, S. X. Yin, L. Liang, Z. Zhou, J. Niu, L.-H. Li, Y. Wang, J. Su and J. Li, *Science* 2018, 361, 48–52.

143. F. J. Uribe-Romo, C. J. Doonan, H. Furukawa, K. Oisaki and O. M. Yaghi, *Journal of the American Chemical Society* 2011, 133, 11478–11481.

144. K. T. Jackson, T. E. Reich and H. M. El-Kaderi, *Chemical Communications* 2012, 48, 8823–8825.

145. Q. Fang, Z. Zhuang, S. Gu, R. B. Kaspar, J. Zheng, J. Wang, S. Qiu and Y. Yan, *Nature Communications* 2014, 5, 4503.

146. H. L. Qian, F. L. Meng, C. X. Yang and X. P. Yan, *Angewandte Chemie International Edition* 2020, 59, 17607–17613.

147. X. Zhuang, W. Zhao, F. Zhang, Y. Cao, F. Liu, S. Bi and X. Feng, *Polymer Chemistry* 2016, 7, 4176–4181.

148. X. Guan, H. Li, Y. Ma, M. Xue, Q. Fang, Y. Yan, V. Valtchev and S. Qiu, *Nat. Chem.* 2019, 11, 587.

149. O. Yahiaoui, A. N. Fitch, F. Hoffmann, M. Fröba, A. Thomas and J. Roeser, *Journal of the American Chemical Society* 2018, 140, 5330–5333.

150. (a) G. H. Bertrand, V. K. Michaelis, T.-C. Ong, R. G. Griffin and M. Dincă, *Proceedings of the National Academy of Sciences* 2013, 110, 4923–4928.

(b) N. Huang, L. Zhai, D. E. Coupry, M. A. Addicoat, K. Okushita, K. Nishimura, T. Heine and D. Jiang, *Nature Communications* 2016, 7, 12325.

(c) P. Pachfule, A. Acharjya, J. Roeser, R. P. Sivasankaran, M.-Y. Ye, A. Brückner, J. Schmidt and A. Thomas, *Chemical Science* 2019, 10, 8316–8322.

151. (a) F. Haase, K. Gottschling, L. Stegbauer, L. Germann, R. Gutzler, V. Duppel, V. Vyas, K. Kern, R. Dinnebier and B. Lotsch, *Materials Chemistry Frontiers* 2017, 1, 1354–1361.

(b) H. Li, P. Shao, S. Chen, G. Li, X. Feng, X. Chen, H.-J. Zhang, J. Lin and Y.-B. Jiang, *Journal of the American Chemical Society* 2020, 142, 3712–3717.

(c) X. Li, Q. Gao, J. Aneesh, H.-S. Xu, Z. Chen, W. Tang, C. Liu, X. Shi, K. Adarsh and Y. Lu, *Chemistry of Materials* 2018, 30, 5743–5749.

(d) D. Bessinger, L. Ascherl, F. Auras and T. Bein, *J. Am. Chem. Soc.* 2017, 139, 12035.

(e) Y. Xiong, Q. Liao, Z. Huang, X. Huang, C. Ke, H. Zhu, C. Dong, H. Wang, K. Xi and P. Zhan, *Advanced Materials* 2020, 32, 2070070.

(f) H. Lv, R. Sa, P. Li, D. Yuan, X. Wang and R. Wang, *Science China Chemistry* 2020, 63, 1289–1294.

(g) K. Lei, D. Wang, L. Ye, M. Kou, Y. Deng, Z. Ma, L. Wang and Y. Kong, *ChemSusChem* 2020, 13, 1725–1729.

(h) Z. Li, Y. Zhi, P. Shao, H. Xia, G. Li, X. Feng, X. Chen, Z. Shi and X. Liu, *Applied Catalysis B: Environmental* 2019, 245, 334–342.

(i) H. Liu, C. Li, H. Li, Y. Ren, J. Chen, J. Tang and Q. Yang, *ACS Applied Materials & Interfaces* 2020, 12, 20354–20365.

(j) Z. Li, S. Han, C. Li, P. Shao, H. Xia, H. Li, X. Chen, X. Feng and X. Liu, *Journal of Materials Chemistry A* 2020, 8, 8706–8715.

(k) Q. Liao, W. Xu, X. Huang, C. Ke, Q. Zhang, K. Xi and J. Xie, *Science China Chemistry* 2020, 63, 707–714.

(l) W. Chen, Z. Yang, Z. Xie, Y. Li, X. Yu, F. Lu and L. Chen, *Journal of Materials Chemistry A* 2019, 7, 998–1004.

152. Y. Xiang, W. Dong, P. Wang, S. Wang, X. Ding, F. Ichihara, Z. Wang, Y. Wada, S. Jin and Y. Weng, *Applied Catalysis B: Environmental* 2020, 274, 119096.

153. (a) S. Li, L. Li, Y. Li, L. Dai, C. Liu, Y. Liu, J. Li, J. Lv, P. Li and B. Wang, *ACS Catalysis* 2020, 10, 8717–8726.

(b) C. Mo, M. Yang, F. Sun, J. Jian, L. Zhong, Z. Fang, J. Feng and D. Yu, *Advanced Science* 2020, 7, 1902988.

154. (a) J. Zhao, J. Ren, G. Zhang, Z. Zhao, S. Liu, W. Zhang and L. Chen, *Chemistry–A European Journal* 2021, 27, 10781–10797.

(b) S. Xu, H. Sun, M. Addicoat, B. P. Biswal, F. He, S. Park, S. Paasch, T. Zhang, W. Sheng and E. Brunner, *Advanced Materials* 2021, 33, 2006274.

155. S.-Y. Ding and W. Wang, *Chemical Society Reviews* 2013, 42, 548–568.

156. B. P. Biswal, S. Chandra, S. Kandambeth, B. Lukose, T. Heine and R. Banerjee, *Journal of the American Chemical Society* 2013, 135, 5328–5331.

157. L. Chen, J. Du, W. Zhou, H. Shen, L. Tan, C. Zhou and L. Dong, *Chemistry–An Asian Journal* 2020, 15, 3421–3427.

158. R. Luo, Y. Yang, K. Chen, X. Liu, M. Chen, W. Xu, B. Liu, H. Ji and Y. Fang, *Journal of Materials Chemistry A* 2021, 9, 20941–20956.

159. C. Qian, Q.-Y. Qi, G.-F. Jiang, F.-Z. Cui, Y. Tian and X. Zhao, *Journal of the American Chemical Society* 2017, 139, 6736–6743.

160. Y. Li, Q. Chen, T. Xu, Z. Xie, J. Liu, X. Yu, S. Ma, T. Qin and L. Chen, *Journal of the American Chemical Society* 2019, 141, 13822–13828.

161. M. Zhang, L. Li, Q. Lin, M. Tang, Y. Wu and C. Ke, *Journal of the American Chemical Society* 2019, 141, 5154–5158.

162. J. W. Colson, A. R. Woll, A. Mukherjee, M. P. Levendorf, E. L. Spitler, V. B. Shields, M. G. Spencer, J. Park and W. R. Dichtel, *Science* 2011, 332, 228–231.

163. D. W. Burke, C. Sun, I. Castano, N. C. Flanders, A. M. Evans, E. Vitaku, D. C. McLeod, R. H. Lambeth, L. X. Chen and N. C. Gianneschi, *Angewandte Chemie* 2020, 132, 5203–5209.

164. W. Chen, L. Du and C. Wu in Chapter 7 - *Hydrothermal Synthesis of MOFs*, (Ed. M. Mozafari), Woodhead Publishing, 2020, pp. 141–157.

165. X. Wang, N. Yang, Q. Li, F. He, Y. Yang, B. Wu, J. Chu, A. Zhou and S. Xiong, *Journal of Solid State Chemistry* 2019, 277, 575–586.

166. (a) Z. Zhao, H. Li, K. Zhao, L. Wang and X. Gao, *Chemical Engineering Journal* 2022, 428, 131006.

(b) J. Klinowski, F. A. A. Paz, P. Silva and J. Rocha, *Dalton Transactions* 2011, 40, 321–330.

(c) R. Vakili, S. Xu, N. Al-Janabi, P. Gorgojo, S. M. Holmes and X. Fan, *Microporous and Mesoporous Materials* 2018, 260, 45–53.

167. (a) K. Yu, Y.-R. Lee, J. Y. Seo, K.-Y. Baek, Y.-M. Chung and W.-S. Ahn, *Microporous and Mesoporous Materials* 2021, 316, 110985.

(b) J. Othong, J. Boonmak, V. Promarak, F. Kielar and S. Youngme, *ACS Applied Materials & Interfaces* 2019, 11, 44421–44429.

168. (a) A. Ghoorchian, A. Afkhami, T. Madrakian and M. Ahmadi, *Electrochemical Synthesis of MOFs*, Elsevier, 2020, pp. 177–195.

(b) V. M. V and G. Nageswaran, *Journal of The Electrochemical Society* 2020, 167, 155527.

169. (a) M. Shahmirzaee, A. Hemmati-Sarapardeh, M. M. Husein, M. Schaffie and M. Ranjbar, *Microporous and Mesoporous Materials* 2020, 307, 110463.

(b) M. Shahmirzaee, A. Hemmati-Sarapardeh, M. M. Husein, M. Schaffie and M. Ranjbar, *ACS Omega* 2022, 7, 3700–3712.

170. (a) B. Szczęśniak, S. Borysiuk, J. Choma and M. Jaroniec, *Materials Horizons* 2020, 7, 1457–1473.

(b) D. Chen, J. Zhao, P. Zhang and S. Dai, *Polyhedron* 2019, 162, 59–64.

171. L. Sun, C. H. Hendon, M. A. Minier, A. Walsh and M. Dincă, *J. Am. Chem. Soc.* 2015, 137, 6164.

172. (a) A. A. Talin, A. Centrone, A. C. Ford, M. E. Foster, V. Stavila, P. Haney, R. A. Kinney, V. Szalai, F. El Gabaly and H. P. Yoon, *Science* 2014, 343, 66–69.

(b) S. Han, S. C. Warren, S. M. Yoon, C. D. Malliakas, X. Hou, Y. Wei, M. G. Kanatzidis and B. A. Grzybowski, *Journal of the American Chemical Society* 2015, 137, 8169–8175.

173. K. Lin, Z. Wang, Z. Hu, P. Luo, X. Yang, X. Zhang, M. Rafiq, F. Huang and Y. Cao, *Journal of Materials Chemistry A* 2019, 7, 19087–19093.

174. C. Krishnaraj, H. S. Jena, K. Leus and P. Van Der Voort, *Green Chemistry* 2020, 22, 1038–1071.

175. S. Chandra, D. Roy Chowdhury, M. Addicoat, T. Heine, A. Paul and R. Banerjee, *Chem. Mater.* 2017, 29, 2074.

176. T. Sick, A. G. Hufnagel, J. Kampmann, I. Kondofersky, M. Calik, J. M. Rotter, A. Evans, M. Döblinger, S. Herbert and K. Peters, *J. Am. Chem. Soc.* 2018, 140, 2085.

177. R. Abazari and A. R. Mahjoub, *Inorganic Chemistry* 2018, 57, 2529–2545.

178. (a) T. Wu, T. Wang, L. Sun, K. Deng, W. Deng and R. Lu, *ChemistrySelect* 2017, 2, 4533–4537.

(b) S. Bhattacharyya, D. Samanta, S. Roy, V. P. Haveri Radhakantha and T. K. Maji, *ACS Applied Materials & Interfaces* 2019, 11, 5455–5461.

179. W. J. Fan, Y. L. Zhao, Y. C. Hu, H. Shi, D. Z. Tan and R. Q. Zhang, *International Journal of Quantum Chemistry* 2018, 118, e25492.

180. C. Chen, C. Tang, W. Xu, Y. Li and L. Xu, *Physical Chemistry Chemical Physics* 2018, 20, 9536–9542.

181. M. Souto and D. F. Perepichka, *Journal of Materials Chemistry C* 2021, 9, 10668–10676.

182. H. Li, J. Chang, S. Li, X. Guan, D. Li, C. Li, L. Tang, M. Xue, Y. Yan and V. Valtchev, *Journal of the American Chemical Society* 2019, 141, 13324–13329.

183. S. Rager, A. C. Jakowetz, B. Gole, F. Beuerle, D. D. Medina and T. Bein, *Chemistry of Materials* 2019, 31, 2707–2712.

184. L. Wang, C. Zeng, H. Xu, P. Yin, D. Chen, J. Deng, M. Li, N. Zheng, C. Gu and Y. Ma, *Chem. Sci.* 2019, 10, 1023.

185. M. Martínez-Abadía, C. T. Stoppiello, K. Strutynski, B. Lerma-Berlanga, C. Martí-Gastaldo, A. Saeki, M. Melle-Franco, A. N. Khlobystov and A. Mateo-Alonso, *Journal of the American Chemical Society* 2019, 141, 14403–14410.

186. N. Huang, K. H. Lee, Y. Yue, X. Xu, S. Irle, Q. Jiang and D. Jiang, *Angewandte Chemie* 2020, 132, 16730–16736.

187. J. M. Rotter, R. Guntermann, M. Auth, A. Mähringer, A. Sperlich, V. Dyakonov, D. D. Medina and T. Bein, *Chemical Science* 2020, 11, 12843–12853.

188. V. Lakshmi, C.-H. Liu, M. Rajeswara Rao, Y. Chen, Y. Fang, A. Dadvand, E. Hamzehpoor, Y. Sakai-Otsuka, R. S. Stein and D. F. Perepichka, *Journal of the American Chemical Society* 2020, 142, 2155–2160.

189. S. Duhovic and M. Dinca, *Chemistry of Materials* 2015, 27, 5487–5490.

190. S. Bi, C. Yang, W. Zhang, J. Xu, L. Liu, D. Wu, X. Wang, Y. Han, Q. Liang and F. Zhang, *Nat. Commun.* 2019, 10, 2467.

191. S. Wei, F. Zhang, W. Zhang, P. Qiang, K. Yu, X. Fu, D. Wu, S. Bi and F. Zhang, *J. Am. Chem. Soc.* 2019, 141, 14272.

192. (a) L. Sun, M. G. Campbell and M. Dincă, *Angew. Chem., Int. Ed.* 2016, 55, 3566.

(b) L. S. Xie, G. Skorupskii and M. Dincă, *Chemical Reviews* 2020, 120, 8536–8580.

(c) L. S. Xie, G. Skorupskii and M. Dincă, *Chemical Reviews* 2020, 120, 8536–8580.

193. M. Souto, K. Strutyński, M. Melle-Franco and J. Rocha, *Chemistry – A European Journal* 2020, 26, 10912–10935.

194. Y. Wu, D. Yan, Z. Zhang, M. M. Matsushita and K. Awaga, *ACS Applied Materials & Interfaces* 2019, 11, 7661–7665.

195. L. Frey, J. F. Pöhls, M. Hennemann, A. Mähringer, S. Reuter, T. Clark, R. T. Weitz and D. D. Medina, *Advanced Functional Materials* 2022, 32, 2205949.

196. Z. Meng, R. M. Stolz and K. A. Mirica, *Journal of the American Chemical Society* 2019, 141, 11929–11937.

197. S. Subudhi, S. P. Tripathy and K. Parida, *Inorganic Chemistry Frontiers* 2021, 8, 1619–1636.

198. P. Lou, L. Li, Y. Wu, L. Yue, Y. Wang, S. Tang, W. Zhang and Y. Cao, *Ionics* 2023.

199. P. I. Scheurle, A. Mähringer, A. C. Jakowetz, P. Hosseini, A. F. Richter, G. Wittstock, D. D. Medina and T. Bein, *Nanoscale* 2019, 11, 20949–20955.

200. M. H. Hassan, R. R. Haikal, T. Hashem, J. Rinck, F. Koeniger, P. Thissen, S. Heiβler, C. Wöll and M. H. Alkordi, *ACS Applied Materials & Interfaces* 2019, 11, 6442–6447.

201. C. Schneider, D. Ukaj, R. Koerver, A. A. Talin, G. Kieslich, S. P. Pujari, H. Zuilhof, J. Janek, M. D. Allendorf and R. A. Fischer, *Chemical Science* 2018, 9, 7405–7412.

202. C.-W. Kung, K. Otake, C. T. Buru, S. Goswami, Y. Cui, J. T. Hupp, A. M. Spokoyny and O. K. Farha, *Journal of the American Chemical Society* 2018, 140, 3871–3875.

203. M. Souto, J. Calbo, S. Mañas-Valero, A. Walsh and G. Mínguez Espallargas, *Beilstein Journal of Nanotechnology* 2019, 10, 1883–1893.

204. L. Sun, C. H. Hendon and M. Dincă, *Dalton Transactions* 2018, 47, 11739–11743.

205. M. M. Samy, M. G. Mohamed, S. U. Sharma, S. V. Chaganti, J.-T. Lee and S.-W. Kuo, *Journal of the Taiwan Institute of Chemical Engineers* 2023, 104750.

206. W. Wang, Y. Zhang, L. Chen, H. Chen, S. Hu, Q. Li, H. Liu and S. Qiao, *Polymer Chemistry* 2021, 12, 650–659.

Index